PUBLIC HEALTH POLICY AND ETHICS

PUBLIC HEALTH POLICY AND ETHICS

Edited by

Michael Boylan

Professor of Philosophy,
Marymount University, Arlington, U.S.A.

KLUWER ACADEMIC PUBLISHERS

DORDRECHT / BOSTON / LONDON

A C.I.P. Catalogue record for this book is available from the Library of Congress.

ISBN 1-4020-2207-7 (e-book)

Published by Kluwer Academic Publishers,
P.O. Box 17, 3300 AA Dordrecht, The Netherlands.

Sold and distributed in North, Central and South America
by Kluwer Academic Publishers,
101 Philip Drive, Norwell, MA 02061, U.S.A.

In all other countries, sold and distributed
by Kluwer Academic Publishers,
P.O. Box 322, 3300 AH Dordrecht, The Netherlands.

Printed on acid-free paper

Printed in the Netherlands.

This volume is dedicated to David C. Thomasma, *in memoriam*.

TABLE OF CONTENTS

PREFACE: WHAT IS PUBLIC HEALTH?

The U.S. based Institute of Medicine has set out the domain of public health as dealing with: epidemiology, health promotion and education, public health administration, international health, maternal and child health, biostatistics, environmental health, and nutrition.[1] No single volume dealing with public health could hope to deal with all of these. In fact, it is the general character of this book to expand the purview even further. In order to get some clarity and focus in this regard, this volume will seek to approach public health first by establishing the theoretical, moral basis for public health, then by examining the nature and structures of certain policies balancing the needs of individuals and society, and finally getting to the practice of public health from a panoramic lens aimed at the landscape already claimed. The end result aspires to infuse a progressive clarity of vision in order to garner support for the expanded topology of public health.

THE HISTORY OF PUBLIC HEALTH

An individual's concern for her own health is as old as conscious humanity. The concern for the health of others when "others" means one's own family or clan must be just as old (for "others" here is virtually synonymous with "self"). What is interesting is when health includes others outside the clan. For example, when Noah is told by God, "I have determined to make an end of all flesh, for the earth is filled with violence, because of them; now I am going to destroy them along with the earth."[2] Noah just gets to the business of building an ark. There is no real compassion shown by Noah for the rest of humanity who would soon die. However, when the issue of Sodom's destruction is on the table, Lot (Abraham's brother) declares, "I beg you, my brothers, do not act so wickedly."[3] Lot's plea to the people of Sodom was made so that they might be saved from the promised destruction. In this way, Lot demonstrates a concern for the health of others (in a fashion more pronounced than Noah). Unfortunately, Lot's call went unheeded so Lot and Abraham gathered up their clan and prepared to depart. In this case the invocation of Public Health fails, but the clan survives—except for Lot's wife who was too concerned about what was happening to others. She ended up as a pile of salt.

Then there is Joseph who helped the Egyptian pharaoh stave off famine by adopting a public policy that created a store of grain in times of plenty so that an impending famine (in which many might starve) could be avoided.[4] The public policy was effective. This is a true instance of effective public health.

But not all Egyptian pharaohs were so concerned with public health. It took nine plagues before the pharaoh at the time of Moses was willing to release the oppressed Israelites.[5] Power and political might were more important than his people's health. It wasn't until the pharaoh's own child[6] was killed that the ruler relented (at first). This is an instance of a prudential grounding of public health.

In Plato's *Republic* the whole state's health was a mirror of individual health: a balance of the classes of people as per the balance of an individual's soul. This may be understood as an exhortation for public health. Plato's holistic vision linked health and happiness with a virtuous life within one's own limits.

Among ancient biomedical writers the Hippocratic writers and Galen were particularly interested in the common causes of disease. In the Hippocratic writers this concern takes the form of very general treatises such as *Airs, Waters, and Places* and more clinical works such as the *Epidemics*.[7] Though there are no clear accounts of Hippocrates engaging in public health practice (as such), Galen was a key physician of the Roman emperor and was called upon for treatment and prevention of disease among those in the army as well as in ancient Rome.[8] The concern among the Romans was prudential: a well-run army to thwart invaders and a healthy capital capable to governing the empire.

For most of human history public health has been reactive. In the ancient world and the Middle Ages there were few preventative measures taken. It was generally a case of reacting to plagues or natural disasters. Few structural fixed programs were maintained except the alienation of lepers from the general population that in some societies occurred for a thousand years. During the Enlightenment, it has been argued that Johann Peter Frank, Benjamin Rush and John Gregory advocated physicians to take a public view of medicine.[9]

The major change in public health occurred in the nineteenth century. Two significant events during this time were: (a) the discovery of the microbial causation of infection, and (b) the recognition that these causal agents could be controlled via measures of sewage treatment, clean water and food, and the use of quarantine for infected individuals.[10]

The understanding of some of the causal mechanisms of disease created a new efficacy for public health that has extended to the present. By knowing underlying causes, the strategies for public health become more efficient. There is a greater confidence that measures taken for prevention of disease or to control the outbreak of infectious disease will be efficacious. It is also the case that some preventative measures, such as vaccination and sanitation, may actually work to protect the population against several sorts of microbial threats.

Today, we are in this last phase of public health. It is more linked to the bio-chemistry of disease. But there are some who think that we should not lose some of the emphases upon balance and life-style that characterized earlier periods. The future of public health may very well be in uniting these emphases in an effort to pro-tect ourselves positively (by the possession of goods that seem necessary to lead an effective life) and negatively (by the prevention of sickness and accident). In the process the rights of the individual often come into conflict with the good of the group. This tension frames much of public health discourse. Within this volume pub-lic health ethics are explored from a number of vantage points (individual vs. com-munity) and our scope ranges far beyond disease to 'health' more broadly understood.

THIS VOLUME[11]

This volume begins in the introduction with a discussion of the grounds of public health. The dichotomy is set between a prudential grounding and a moral grounding (with the latter being the preferred view). Next, in chapter one, Micah Hester sets the current stage of public health discussion with a crucial problem of balancing the needs of the group vs. those of the individual. This is a question of justice that must be addressed in order for effective public health policy to go forward.

In chapter two Edmund Pellegrino and the late David C. Thomasma extend Hester's concerns and argue that an effective ethic of social medicine might be con-structed by extending (by analogy) the existing ethic of clinical medicine. In order to accommodate this balance it is necessary to create a philosophy of society based upon the four hierarchical social ends of medicine within this shared community world-view. In this way Pellegrino and Thomasma (using Aristotle and Aquinas as their guides) extend and give specification to the moral grounding of public health advo-cated in the introduction.

In chapter three, Rosemarie Tong depicts conflicts between the approaches and ends of clinical practice and those of public health. The two can work in different directions. As she harkens back to some of the points raised by Pellegrino and Thomasma, Tong begins her own holistic analysis in order to balance the risks and rewards of implementing a public health response (as opposed to individual clinical responses) to the problem of obesity in America. Her nuanced analysis offers not only a response, but a way of thinking about such problems from a systemic vantage point (cf. also Teays and Purdy).

In Deryck Beyleveld and Shaun Pattinson, chapter four, the focus becomes the individual and his legitimate rights claims (cf. Green and Cummiskey). Using Alan Gewirth's Principle of Generic Consistency as their guide, this essay explores the problem of allocation of some sorts of medical treatments. In certain cases, the balance of justified individual claims creates equality in distribution and some restraints upon medical research.

Justice is also the topic of chapter five. Rosamond Rhodes uses the public health frame of triage to examine allocation of care in cases of terrorism. Again, the reader is presented the problem of the rights of the individual as opposed to a group solution approach. This contrast is made through examining John Rawls' principles of "fair equality of opportunity" and the "difference principle" (as amplified by Norman Daniels) to illustrate the position of prioritarianism, against that of utilitarianism. Because the frame of the essay is triage, Rhodes argues for what she believes are reasonable principles to guide allocation of resources that are group oriented (with certain safeguards).

In chapter six, Jacquelyn Kegley extends the discussion of the group vs. the individual through an interesting twist: genetics. In genetics both the character of the individual and the group are affected. Automatically, both perspectives are brought to the fore. This self-referential mapping may involve a revolutionary change in perspective. As a result of this change, the individual and the group must be addressed in a new way—via their mutual interrelationships. This relationalism offers a novel response to the group vs. the individual dilemma.

Chapter seven examines a number of conflicts between the priorities of the public view with that of the individual. In this instance the individual has a disability (*fetal spina bifida*). Using a dialectical approach Mary Mahowald plays the various arguments against each other to bring into relief the overlapping qualities of each (e.g., quality of life vs. sanctity of life or public priorities vs. personal needs). This dialectic is put forth to direct a public moral discussion (consistent with the dialectical method suggested in the introductory essay).

The use of dialectic continues with my essay on gun control, chapter eight. In this essay the various positions of the gun advocates are matched against those opposed to gun ownership. This dialectic is then set into the public realm as an issue that transcends individual preference.

Just as gun control often slips beneath the public radar, so also is the even more pervasive incidence of domestic violence. Wanda Teays addresses this public health disaster in chapter nine. Like most public health epidemics, the problem of domestic violence will continue to get worse unless specific action is taken. However, in this case the requisite action may be macro in scope. Just as Tong pointed to systemic problems involved with obesity, Teays sees the causes of domestic violence as systemic, as well. It will do little good to depend upon isolated clinical responses if we want to turn the tide. Instead, a series of systemic remedies aimed at existential and institutional factors are necessary in addressing this public health tragedy.

Chapter ten also addresses the systemic dimensions of the causes of ill health that may lead to disability and death. In a bold turn Laura Purdy focuses internally upon the profession of bioethics, itself. Can bioethicists talking about public health really be part of the problem? Certainly, if they become too narrowly, clinically focused to the exclusion of broader, holistic concerns that may involve deep-set social and political realities. Like those advocating a systemic vision, Purdy argues for a broad perspective that includes discussion of governmental policy and social institutions. Such a position is consonant with the U.S. Institute of Medicine that sees environmental health (broadly conceived) as a legitimate concern of public health.

David Cummiskey in chapter eleven employs a methodological perspective that has some affinity to Beyleveld and Pattinson, viz., viewing public health responsibility from the vantage point of the individual and his dual right to die and to health care. In the case of Cummiskey, he is grounding both rights in a Kantian-style understanding of autonomy.[12] If the foundation principle is adopted (conferring a pattern of rights claims), then it doesn't matter if its consequential application brings surprising results.

The issue of health care rights continues with Michael Green in chapter twelve. Green employs an examination of the methodology of Henry Shue in order to ascertain whether health care is a defensible human right, and if so, to what extent may it be applied and whether are there any correlative duties? In his exposition, Green follows alongside Cummiskey and Beyleveld and Pattinson in answering the group problem via an exposition of the perspective of the individual and then universalizing.

The final essay in the collection, chapter thirteen, by Rosemary Quigley responds to many of the concerns raised in the book: first the balance between the individual and the group, but also, the means for ascertaining the interests of each and how they can become recognized. One policy initiative that seeks to do this is the creation of advocates. This seems as if it might be the perfect solution: an individual with sympathies for the group acting on behalf of the group. However, this mechanism is also beset with problems. In the situation with AIDS, it was the efforts of affected individuals that gave authentic voice to their concerns. Also, there is the problem of objectivity and advocacy. In the end, Quigley brings to focus the complexity of consequences and suggests safeguards for the individual and parameters for the way that advocacy should operate.

ACKNOWLEDGEMENTS

As in any project of this dimension there are many to thank. I would like to begin by thanking David C. Thomasma, *in memoriam,* to whom this volume is dedicated. Dr. Thomasma was the individual who originally accepted this book to be in his series. He was also a pioneer in the emerging field of bioethics. He was a man of great intelligence and goodness.

Second, I'm grateful to Thomasine Kushner who put the idea into my head to edit a volume on public health ethics in the first-place. Thanks Tomi.

Next, I would like to thank Marymount University for giving me two grants to work on and complete this book. I would like to mention, in particular, Dr. Loretta Seigley (vice-president of academic affairs), for her support of this project.

I would next like to extend my gratitude to all the contributors in this volume who have cheerfully met deadlines and accommodated my various comments and requests.

Then there are the two anonymous reviewers whose suggestions improved the volume. I would also like to thank the crew at Kluwer: Anne M. Ultee, senior publishing editor; Jacqueline Bergsma, publishing editor; and Helen Van der Stelt, assistant to the publishing editor.

This book would not have happened without the production assistance of Matt Dunham. Matt's technical support was essential!

Finally, I should mention the continuing support and love of my family: Rebecca, Arianne, Seán, and Éamon. They constitute a micro community in which I've been engaged in public health issues for years!

[1] Institute of Medicine, Committee for the Study of the Future of Pubic Health. *The Future of Public Health.* (Washington, DC: National Academy Press, 1988, pp. 225f.

[2] Bruce M. Metzer and Roland E. Murphy, eds. *The New Oxford Annotated Bible* (NY: Oxford University Press, 1994), *Genesis* 6:13.

[3] *Genesis* 19:7.

[4] *Genesis* 41:15-36.

[5] *Exodus* 7-10.

[6] *Exodus* 12:29

[7] E. Littré, *Oeuvres complète d'hippocrate* 10 vols. (Paris: J. B. Baillière, 1851). Standard edition.

Corpus Medicorum Graecorum (Berlin: Akademie-Verlag, on-going). New editions of selected texts with excellent notes and apparati by various editors.

[8] See: Owsei Temkin, *Galenism: The Rise and Decline of a Medical Philosophy.* (Ithaca, NY: Cornell University Press, 1973); Vivian Nutton, "Galen *ad multos annos*" *Dynamis* 15 (1995): 25-39. Wesley Smith, *The Hippocratic Tradition.* (Ithaca, NY: Cornell University Press, 1979).

[9] Steven S. Couglin and Tom L. Beauchamp, eds, *Ethics and Epidemiology* (NY: Oxford University Press, 1996).

[10] These causal factors are discussed by Peter J. Lachmann, "Public Health and Bioethics" *Journal of Medicine and Philosophy* 23.3 (1998): 297-302.

[11] I should note that this section is intended to suggest the interrelationships between essays and is not merely a series of summaries.

[12] Of course Beyleveld and Pattinson argue from the position of Alan Gewirth, however; there are many fundamental similarities between these standpoints see: Deryck Beyleveld, "Gewirth and Kant on Justifying the Supreme Principle of Morality" in Michael Boylan, ed. *Gewirth: Critical Essays on Action, Rationality, and Community* (Lanham, MD: Rowman and Littlefield, 1999), pp. 97-118.

MICHAEL BOYLAN

INTRODUCTION

The Moral Imperative to Maintain Public Health

ABSTRACT: This essay examines two candidates for the grounding of public health: (a) prudential grounds and (b) moral grounds. It is argued that a prudential grounding for public health is inadequate because such a worldview is inconsistent and violates the Personal Worldview Imperative. Therefore, a moral grounding for public health is put forth that satisfies conditions of thinking of a nation as a collection of healthy individuals and of thinking of a nation as a macro unit that is balanced. To this end, human rights and justice are important ingredients.

KEY WORDS: public health and ethics, public health and morality, worldview, human rights.

There are at least two sorts of imperatives to maintain public health[1]: (a) prudential, and (b) moral. In the former case an agent advocates policies supporting public health because it makes the environment in which the agent lives more desirable for the agent, himself. In this case (for example), one would like to rid his area of cholera because if cholera is allowed to spread, then the agent, himself, might catch cholera. In such situations the agent is thinking only of his own advantage. This has two discernable effects: (1) public health is merely an extension of particular agents' own personal needs, and (2) [as a result of #1] public health policies will only be supported when there is a political mandate to do so based upon coalitions of people advocating their shared self-interest.

Should public health be merely a line item in the intercourse of self-interested policy maneuvering or should it be governed by some other sort of principle? This essay will discuss this first alternative and demonstrate its shortcomings. Then it will propose a different ground of public health policy based upon a worldview model and fundamental human rights.

PRUDENTIAL GROUNDS FOR PUBLIC HEALTH

The prudential model is based upon a principle of selfish egoism and extended egoism (the political expression of selfish egoism). Continuing with the cholera example, agents are only after their own self-interest. Thus, these agents will support a policy if and only if they believe that policy will directly benefit them. In the cholera example, someone living outside a city might only support a public health effort for sanitation if he believed the danger of contracting cholera was getting dangerously close to infecting him or his family. The risk must be immanent. There must be a clear and present danger of his being harmed. This sort of agent is willing to support only those projects that directly benefit him. From the agent's point of view, this is the most efficient allocation of resources. ("Efficient" here means not spending public money on other people apart from the agent. Thus the adage, *if the program doesn't help me, it's a wasteful program.*)

Others will support policies that they see it in their "enlightened self-interest." These might include preventative measures that may (indirectly) help others whether or not there is an immanent threat or a clear and present danger. These individuals are acting from self-interest but have a longer view of things. They see prevention as the most efficient allocation of resources because reacting in the midst of a crisis is notoriously expensive. These individuals would point to the adage, *an ounce of prevention is worth a pound of cure.* In this way the "enlightened self-interest" version of egoism sees public health measures as some sort of insurance policy that will efficiently address potential problems. ("Efficiency" here means using fewer public dollars to address an issue that may have an impact upon the agent, himself. Though there is some waste involved because (i) the problem might not arise and (ii) the solution may help many others apart from the agent—still, the cost savings from acting early offsets this other sense of waste.)

These two strategies can be summarized in the following figure.

Direct Self-Interest Version

Condition	Response	Justification
Cholera about to infect One's own locale	Sanitation of one's own locale	Agent is interested in his own and his family's well-being

"Enlightened Self-Interest" Version

Condition	**Response**	**Justification**
Cholera has infected Many cities in the area	Preventative measures taken locally or regionally, or nationally	Even if the preventative measures help more than the agent, they are cost effective insurance for the agent

Figure One: The Direct and "Enlightened Self-Interest" Versions of Prudential Justification of Public Health Policies via Extended Egoism

Though these two sorts of prudential justification share the category of extended egoism, they otherwise are quite distinct. In the direct form, the agent views "efficiency" solely in terms of how it benefits himself. In other words, all money spent upon others is "wasteful." This is because only the agent, himself, counts. If a public program does not benefit him directly—right now—it is wasteful.

To some (including the author) this view of policy wastefulness is problematic. The reason I would give for this is that it violates the Personal Worldview Imperative: "All people must develop a single comprehensive and internally coherent worldview that is good and that we strive to act out in our daily lives."[2] Direct egoists are forced to adopt an inconsistent position of creating a worldview in which they count for more than others. But on what grounds is this assertion put forth? Merely that the agent sees and values her own claims for goods above all others. But what is the basis of this valuation? Simply that the agent is self-absorbed to the extent that she dismisses other people's claims as being of lesser weight because they are not the agent's own claims. But why this inequality in claim valuation? The basis is centered in the original assessment by the agent that she is more important than other agents merely because she says so. However, if other agents make the same claim, then a contradiction develops immediately. (Logical contradictions within individual worldviews violate the Personal Worldview Imperative.)

1. I, Ms. Greed, value my claims for x, y, and z (where x, y, and z, are goods of agency in short supply)—Fact
2. Others also value x, y, and z—Fact
3. The speaker's claims hold positive asymmetric weight over competing rights claims—Assertion (essential to selfish egoism)
4. Ms. Greed is a speaker—F
5. Ms. Greed's claims are more important than any other competing rights claims—1-4
6. Mr. Grasping is among the competing rights claimers—F

7. Mr. Grasping is a speaker—F

8. Mr. Grasping's claims are more important than any other competing rights claims (including Ms. Greed's)—3, 6, 7

9. There can be only one agent whose claims are more important than any other competing rights claims—F

---

10. The formula espoused in premise #3 is logically inconsistent—5, 8, 9

Figure Two: The Logical Inconsistency within the Egoist's Worldview

In the "enlightened self-interest" version of prudential justification via extended egoism the agent understands that it may be more fiscally efficient to create programs that treat problems before they arise (preventative). This is done because it spends less of the public's money and will cause the agent to have lower taxes, etc. This sort of agent views such policies as insurance effectively meted in an efficient social manner. Efficiency here has to do with lower net social costs *but only for the sake of the agent himself*. There is no altruism involved nor connection with the general good. Because the "enlightened" egoist (along with the direct egoist) asymmetrically values his rights claims above others, and because this is an inconsistent position, he is also subject to the same negative evaluation (as per the Personal Worldview Imperative).

Some would assert that the addition of "extended" to egoism is significant. This is because it is thought that if one accepts a policy that extends beyond the self to include others, then it automatically becomes moral. The inclination to this position is most often brought forth by utilitarians.[3] However, this assessment is incorrect. In the case of direct egoists, the only reason that they support extended versions of their vision is to protect themselves. This is a far cry from signing on to the position that one endorses the happiness/pleasure of the group. Endorsing the happiness/pleasure of the group as primary *means* that the individual has renounced the vision that the individual's own happiness is primary (an essential maxim to direct egoists). And yet, it is easy to see why some utilitarians might like to make this claim. This is because 19th century utilitarianism was "sold" on the scientific principle of selfish egoism (to which most in the audience would agree) and then this was extended to the group. The extension to the group constituted the "moral turn" and made a prudential theory into a moral theory. However, this move has been called into question by no less than the prominent utilitarian, Henry Sidgwick, himself.

> I do not mean that if we gave up the hope of attaining a practical solution of this fundamental contradiction [between egoistic and general hedonism] . . .

it would be reasonable for us to abandon morality altogether; but it would seem necessary to abandon the idea of rationalizing it completely. We should doubtless still, not only from self-interest, but through sympathy and sentiments protective of social well-being, imparted by education and sustained by communication with other men, feel a desire for the general observance of rules conducive to general happiness . . . . But in the rarer cases of a recognized conflict between self-interest and duty, practical reason, being divided against itself, would cease to be a motive on either side; the conflict would have to be decided by the comparative preponderance of one or the other of two groups of non-rational impulses.[4]

When put up against it, there is no compelling rational reason why the direct egoist should accept the claims of the general happiness as primary and action guiding. This is because the very foundation of the direct egoist is that he is committed to his own self-advancement above all else. To single one's self out asymmetrically over all others is generally not the worldview of one willing to share and to be cooperative.[5] It is the worldview of the free rider, exploiter, or *kraterist* dedicated to the laws of *feng*.[6] Thus, it seems to this writer that there is no way that the direct egoist might be construed as anything other than a prudential agent bargaining about preferment that he feels is his due.

"Enlightened egoists" fare no better. The fact that they will permit extended policies that do not directly benefit them is *not* a sign of cooperation but rather is indicative of their understanding of what constitutes "efficiency." The enlightened egoist, just like the direct egoist, is only concerned about herself. However, the manner of *effecting her self-interest* is different. The tactics of attaining personal advantage are set in terms of a strategy that spends less total money. But why should the enlightened egoist care about the least amount of social money being spent? Shouldn't she be as the direct egoist and consider efficiency solely in terms of money spent upon herself? No. This is because the more generalized efficiency calculation is based upon a premise that if social funds are used effectively, then either taxes (upon the agent) will be lower or there is more social money left over to spend upon efficient programs to benefit herself. In either case, the enlightened self-interest agent is in no way cooperative or interested in the general happiness, but only seeks a unique strategy of efficiency that differentiates herself from the direct egoist.

If it is true that both direct and enlightened egoists are logically inconsistent in their views (as per Figure Two), and if logically inconsistent views are rejected by the Personal Worldview Imperative, then both of these positions should be rejected as foundations for Public Health.

MORAL GROUNDS FOR PUBLIC HEALTH

How does Morality Relate to Public Health?
In section one of this essay, it has been argued that both the direct and enlightened egoists fall prey to logical inconsistency—even in their extended forms and so are not suitable justifications for public health policy. In its stead, this essay will suggest that moral grounds for public health are more certain because they give a clear and inter-subjective foundation.

To begin, it is essential to address the difficulty presented from the prudential worldview: the inconsistent claims of asymmetrical preference. These may be addressed directly in the following argument.

The Argument for the Moral Status of Basic Goods

1. Before anything else, all people desire to act—Fact
2. Whatever all people desire before anything else is natural to that species—Fact
3. Desiring to act is natural to *homo sapiens*—1,2
4. People value what is natural to them—Assertion
5. What people value they wish to protect—Assertion
6. All people wish to protect their ability to act beyond all else—1,3,4,5
7. The strongest interpersonal "oughts" are expressed via our highest value systems: religion, morality, and aesthetics—Assertion
8. All people must agree, upon pain of logical contradiction, that what is natural and desirable to them individually is natural and desirable to everyone collectively and individually—Assertion
9. Everyone must seek personal protection for her own ability to act via religion, morality, and/or aesthetics—6.7
10. Everyone upon pain of logical contradiction must admit that all other humans will seek personal protection of their ability to act via religion, morality, and/or aesthetics—8,9
11. All people must agree, upon pain of logical contradiction, that since the attribution of the basic goods of agency are predicated generally, that it is inconsistent to assert idiosyncratic preferences—Fact
12. Goods that are claimed through generic predication apply equally to each agent and everyone has a stake in their protection—10,11
13. Rights and duties are correlative—Assertion
14. Everyone has at least a moral right to the basic goods of agency and others in the

society have a duty to provide those goods to all—12, 13

For the purposes of this essay, premise #11 is perhaps the most controversial. This is because it declares that when predication occurs generally, then all instantiations share (via logical heritability) in that general characterization. Thus if Mary is a human and if humans are rational animals, then Mary's claim to being human is true in virtue of "rationality" being heritable from the generic attribution. Mary is rational not because she is Mary, but because she is human. Thus, for Mary or any other agent to assert idiosyncratic preference (the basis of asymmetric claims) is for Mary to make a logical mistake. When making reference to generically predicated properties, it is a logical mistake to declare that they arise from some disposition of the agent, herself. Instead, the agent (upon pain of logical contradiction) must admit that generic features of action are predicated and apply generally. Because of this, humans (with respect to these generically predicated properties) are interchangeable. Each counts the same.

But what are these goods that may be predicated of all people in virtue of being human? This is a difficult question, but one version of these goods is as follows:

The Table of Embeddedness

BASIC GOODS
Level One*: Most Deeply Embedded* (That which is absolutely necessary for Human Action): Food, Clothing, Shelter, Protection from unwarranted bodily harm

Level Two: *Deeply Embedded* (That which is necessary for effective basic action within any given society)
Literacy in the language of the country
Basic mathematical skills
Other fundamental skills necessary to be an effective agent in that country, e.g., in the United States some computer literacy is necessary
Some familiarity with the culture and history of the country in which one lives.
The assurance that those you interact with are not lying to promote their own interests.
The assurance that those you interact with will recognize your human dignity (as per above) and not exploit you as a means only.
Basic human rights such as those listed in the U.S. Bill of Rights and the United Nations Universal Declaration of Human Rights

SECONDARY GOODS

Level One: *Life Enhancing,* Medium to High-Medium Embeddedness
Basic Societal Respect
Equal Opportunity to Compete for the Prudential Goods of Society
Ability to pursue a life plan according to the Personal Worldview Imperative
Ability to participate equally as an agent in the Shared Community Worldview Imperative[7]

Level Two: *Useful,* Medium to low Medium Embeddedness
Ability to utilize one's real and portable property in the manner she chooses
Ability to gain from and exploit the consequences of one's labor regardless of starting point
Ability to pursue goods that are generally owned by most citizens, e.g., in the United States today a telephone, television, and automobile would fit into this class.

Level Three: *Luxurious,* Low Embeddedness
Ability to pursue goods that are pleasant even though they are far removed from action and from the expectations of most citizens within a given country, e.g., in the United States today a European Vacation would fit into this class
Ability to exert one's will so that she might extract a disproportionate share of society's resources for her own use.

What this Table of Embeddedness does is to suggest what goods are comparatively more embedded to action. If the desire to act is the most primary human desire, then the Table of Embeddedness provides a hierarchical ranking of those goods in order of necessity.

It would seem that the recognition of the fundamental character of action (seen generically) is what is meant by morality—that I describe as the science of right and wrong in human action.[8] Under this guise, what generally supports the possibility of human action (and its effective action) must be a primary principle of morality. If this is true, then under a social interpretation of distributive justice, a country must allocate the basic goods (level-one) for all citizens before all else (subject to the caveat of 'ought implies can').[9]

'Public Health' under this interpretation is broadly defined as providing citizens the goods of agency represented in level-one and level-two basic goods and (when possible) level-one secondary goods. This is because 'health' as a social concept is concerned with allowing people to actualize themselves fundamentally as effective acting agents.[10]   Public health is therefore broadly concerned with supplying the goods necessary for human action and eliminating impediments to the same.[11] Such a definition can be interpreted from two angles: (a) a collection of healthy humans, and (b) the maintenance of a healthy society—seen from the perspective of the macro group. From the perspective of (a) it is better to be broader in our scope and include the first three levels of goods because without these people cannot be effective actors in society. If action is most essential to our nature as humans living in a social world, then removing those impediments will provide opportunity for all to exercise their autonomy subject merely to reasonable social constraints (such as those described in the Personal Worldview Imperative). Thus, from the perspective of (a) what is being advocated is a set of policies that will empower as many people as possible to be able to act as they wish in society (subject to reasonable social constraints). Public health is about supporting the individual health of each person (understood as allowing her those goods requisite for effective action).

In sense (b) the concern is with creating a *society* that (seen as a sociological unit) is healthy. Such a society would be one in which no segment of the society is being left out. Thus, if one viewed his society and found an identifiable social group (with robust boundaries) that was being excluded from full and equal participation in the society, then that *society* would be unhealthy. To create a healthy social unit, the application of distributive justice along the lines of the Table of Embeddedness would be an important step toward bringing back society's health. Thus, under interpretation (b) public health is about making sure that the first three levels of the goods of agency are given to all at some appropriate functional level.[12] In this way the state as a social unit may become healthy.

The duties entailed by both the (a) and (b) approaches begin with level-one basic goods, those goods that biologically are requisite for human action. From the public health perspective the lack of these severely limits an agent's ability to act. Without adequate food, children are more likely to become mentally retarded. Adults are likely to develop other diseases and mental illness. Without adequate clothing and shelter people can die. Without proper sanitation for the same, individuals are subject to added external and internal disorders that will severely affect their ability to act. Finally, without reasonable freedom from bodily harm, the agent's attention is turned toward the attacking entity with the result that free deliberation is severely limited. This bodily harm can come in the form of: sickness, accident, crime, war, and domestic abuse (among others). Thus, since all agents have a claim to possess (at least) level-one basic goods, the first responsibility of public health is to provide these (as much as possible)[13] to all agents living within the society. Under this interpretation of public health, homelessness, poverty, disease, common accidents (like carpal tunnel syndrome), crime, civilian war casualties, domestic abuse (among others)[14] are all issues of public health (in both senses).

But the moral obligations of public health do not stop there. Level-two basic goods are also important for effective agency. These include educational goods such as literacy and other educational opportunities. Under this interpretation of public health, illiterate members of a society are "unhealthy" (and so is that society) because some people are relegated to second-class status as members of the underclass. The feelings of being dispossessed that many experience because of their lack of basic, practical education is a public health tragedy.

The dispossessed in every society are also subject to exploitation and loss of fundamental human rights. For example, it is a common belief that crime is higher among the poor.[15] The perpetrators of these crimes are often "profiled" so that the police are more likely to stop anyone who fits a certain profile whether she is guilty or not.[16] Thus, an innocent person of x-racial/ethnic type may be subject to loss of human rights just because she is a member of the x-group.

Though the claims of agents for level-two basic goods are not as strong as those of level-one basic goods, they are still of significant importance for policy consideration.

The final category of goods that I would put forward as of interest in the foundations of public health are the level-one secondary goods. These goods deal with respect, equal opportunity, and a reasonable chance to create a life that the agent believes in (subject to the caveats mentioned in the Personal Worldview Imperative). While these goods are the least embedded of those affecting Public Health, they are nonetheless important. For example, if in India the lowest "untouchable" caste is (*de facto*) denied actual access to preferred jobs in the society (despite being capable), then this sort of lack of equal opportunity diminishes their reasonable expectation of effective action. Again, if certain career paths are closed to individuals because of gender (which occurs explicitly (*de jure*) in some and implicitly (*de facto*) all over the world), then this diminishment of ability to act constitutes a public health crisis of the third order.

Under this account, the hierarchy of public health claims follows the ordering of the Table of Embeddedness from level-one basic goods to level-one secondary goods. (Level-two and level-three secondary goods are agent-driven goods that are sought and attained according to the economic model governing that society and are thus not under the domain of public health responsibility.)

How Should the Moral Imperative affect Public Health Policy?
In the last section it was argued that moral claims based upon the nested goods of
agency provided the scope of public health. This is obviously much wider than some
traditional notions that restrict public health to sanitation and disease control.[17] The
reason for this is that health care concerns are often restricted to issues arising from
disease or accident. These are the traditional reasons that one sees [?] physician. If
what the physician does [?] large [?] what [?] then the account
that has been pr [?] only about the
type-(a) sense of [?] (a collection of
healthy humans [?] upon disease
and accident-pre [?] nomous action
subject to the [?] lively debate.
However, if one [?] a *society* that
(seen as a sociolo [?] ty is deprived
of the goods of [?] public health
must be rejected. [?] the necessary
goods of agency f [?] with the PWI
is more healthy th [?]

But what is th [?] health? The
answer that I woul [?] hat affect the
longevity of life a [?] an many are
willing to admit. [?] ic/economic
group within the U [?] any other country) that is discriminated against her
by virtue of her membership in these various social groups (loss of level-one secondary goods and possibly level-two basic goods), then Juanita's health and her susceptibility to accident is increased.[18] One's individual health is affected by her social environment. If the generating assumption is that fundamentally all humans by nature
desire to purposively act, then denial of any of the essential goods of agency (according to their nested order) will move an individual away from what she fundamentally
is: a purposive agent living in a social world. This is what people naturally are. To
deny people their natural actualization is to make them unnatural. When people are
forced to be unnatural, then it is my conjecture that they will (individually) be more
prone to sickness and accident. Such a group of people (under this assumption) will
be prone to ill-health. Thus, under the type-(a) sense of public health the denial of the
goods of agency will create an unhealthy populace.

Since both definitions of public health (a) and (b) require a broader rather than a
narrower definition, it is incumbent to see what the consequences of this vision are.

What Would a Morally-Based Public Health Policy Look Like?
In this last sub-section of the essay we will briefly explore the broad outlines of a public health policy. To begin, it must be admitted that any country committed to the broader mission of public health as advocated in this essay will have to engage in some sort of rationing formula. This is because not even the richest country in the world could fulfill the allocation of all these goods, in full, to all potential claimants. The very depiction of the goods themselves lends itself to open-ended mandates that would exceed the resources of any country on earth. No country can provide everything. Thus, some sort of allocation formula is called for. This is an instance of distributive justice.

In the first sense of public health (a) discussed above, this allocation formula would view the arena of distribution as a collection of discrete separate individuals. Why should we support one group of goods to John Smith over Rosilita Jimminez? Various traditional (prudential) considerations in the creation of public health policy (type-a) refer to the relative wealth of John Smith and Rosilita Jimminez. The wealthier person can pay for services directly out of her own pocket or influence government directly (via contributions) or indirectly (via membership in a group that has its own group of lobbyists). Thus, under conventional (prudential) rationing mechanisms relative individual wealth plays a key role in the allocation of society's resources to said individuals. This sort of allocation formula refers to one's personal power as the guiding principle for garnering society's resources. This krateristic formula fits in fine with the prudential ground for public health, but it is in conflict with the moral grounds. Since it has been established that moral grounds are more primary to the principles of human action, and that no person has grounds for idiosyncratic preference, it seems clear that the wealthy and the powerful ought not be allowed to garner a disproportionate share of society's resources for public health.[19]

Instead, public policy should follow the Table of Embeddedness in allocating the basic goods of agency before secondary goods according to their level of embeddedness. In the case of there being a conflict of equally embedded goods (such as two individuals competing for a good that will save their respective lives—like an organ transplant), then moral grounds cannot decide the question. Both individuals have an equal moral right to the organ that will save their lives. But who needs the organ the *most*? If patient x seems as if she will die in three days without the organ transplant and patient y seems as if he will die in one day without the transplant, then (on these somewhat simplified criteria alone) patient y should receive the transplant.

But what if there is one organ and both patient x and patient y both need the transplant in order to live to the next day? At this point, there is no moral criterion to choose between them. In such cases, a conspicuously random procedure should be employed. Thus even if x is a millionaire and y is a street person, their lives will count equally in the moral calculus.

The real enemy to this position is the advocate of kraterism. This individual believes that he is *entitled* to more than others because he has *earned it* (even when '*earned it*' only means being born into a particular family line). But even if individual x has made her entire fortune herself, her claim to a distributive public health good is no different than y's. This is because kraterism is merely a prudential allocation formula while the Table of Embeddedness is the foundation of a moral allocation formula.

Thus, under interpretation-a of public health (a collection of healthy humans) moral criteria trump prudential criteria.

The second sense (b) of public health (a healthy society) is, of course, relative to *what* one takes to be a healthy macro unit. Are we to use measures such as: greatest aggregate output, or greatest average productivity? There are many econometric devices to measure prudential success. And in the second sense (b) of public health, prudential success is not without some merit as a factor for consideration. If the society were very inefficient, then it will produce fewer goods to be distributed. Therefore, we cannot ignore the connection between prudential success (from a macro vantage point) and the ability of a society to fulfill its moral mission of public health. However, at the same time, it is not the case that attention to economic efficiency should be the sole or principal distributional mechanism. If it were, then prudential concerns alone should drive public health policy. Instead, I would contend that among economically developed[20] nations, moral concerns should drive public health policy even if this means decreased efficiency. The reason for this is that "health" is taken to mean balance. The only way to achieve balance is to have a counterbalance to simple efficiency as the sole decision making criterion. Efficiency is not to be totally ignored, but the principal guiding principle must be the Table of Embeddedness (aka morality). It is by creating policies that will steer resources (via taxation or other mechanisms) to all citizens (at least) at the level of basic goods that one might be able to judge the society, as a whole, as balanced. The enemy to this vision is the *laissez faire* policy maker who thinks that without interference everything will sort itself out just fine. It is true that without interference everything will sort itself out, but at what cost? One could, for example, say that we should provide health care only to those who can afford health insurance. In the United States, at the writing of this essay, this would leave out 42,554,000 Americans without health insurance and not on public assistance, and 36,100,000 Americans on public assistance with haphazard Medicaid coverage on a five-year time clock. This means that almost 28% of Americans are or could be without recourse to one of the basic goods of agency, health care (aka freedom from unwarranted bodily harm).[21]

In order to cover these individuals through taxes, or some other public mechanism, it would make the United States' economy less efficient. Some level of adequate care[22] should be available to all even if it lowers societal efficiency and the ability of many citizens to own luxury goods (level-three secondary goods).

Thus, under interpretation-b the sociological balance could be measured by a map that depicted the topography of need by various colors: Red for those regions in which the number of people without level-one basic goods is 2.5% or greater;[23] Orange for those regions in which the number of people without level-two basic goods is 2.5% or greater; Blue for those regions in which the number of people without level-one secondary goods is 2.5% or greater; and Green for those areas in which at least 97.5% of the people have all three. Under interpretation-2, it is only when our map is virtually green that we may boast of having created policies that have achieved public health.

CONCLUSION

There are (at least) two ways to view what is public health. The first way is to understand it as a collection of healthy individuals. The second way is to view the society as a unit in itself and to recommend ways of making that macro unit better. In both instances, prudential grounds for policy will lead to a skewed outcome that will not achieve the avowed goal. In the case of the collection of healthy individuals this means that money and influence will provide the goods of public health only to the wealthy (whether they deserve their wealth or not). This violates the moral dictum of viewing each person without idiosyncratic preference.

In the case of the healthy unit, efficiency (as measured by various social science statistics) becomes a crucial player. However, if it is the supreme arbitrator, then the result will be a plutocracy in which only the wealthy will have the essential basic goods of agency. Efficiency is important, but it must give way to moral equity as a distribution formula by which society might provide as many of the basic goods of agency to all. Only this sort of policy orientation will create something approaching balance in the society. And since health is balance, it is only this sort of strategy that will create a healthy society (interpretation-b).

No matter how you understand public health (senses (a) or (b)), it is only by moral means that the enterprise itself makes sense. Idiosyncratic preference (that stands behind the prudential approach) is itself irrational. If rationality is (at least) a necessary condition of authentic human action,[24] then if we pursue prudential concerns when moral imperatives are in conflict and when those moral imperatives are more rationally primary, then we degrade ourselves and we become co-conspirators in devolution of society itself.

Our individual and collective futures lie in not allowing this to happen.

NOTES

[1] It should be noted from the outset that the purview of this essay is public health within some given society. This is not to imply that I do not believe that there are international responsibilities involved in maintaining public health, but merely that the sovereignty structure of the world at present makes this sort of discussion more readily applicable. I do believe that ultimately these responsibilities should extend internationally. The appropriate agencies to administer these programs need to have greater authority than the present U.N. programs or those of the World Health Organization or the World Bank. (The WHO's actions in regard to the recent SARS outbreak is a case in point.) At this moment in history, it seems that the only way to execute duties is first through one's sovereign state and then remotely via these fledgling international bodies.

[2] Michael Boylan, *Basic Ethics* (Upper Saddle River, NJ: Prentice Hall, 2000), p. 27-28.

[3] This has been a much discussed position in utilitarian thinking from the beginning. For a sample of the development of this notion see: I. Historical Sources. John Austin *The Province of Jurisprudence Determined,* ed. by Robert Campbell, 5th edition. 2 vols. (London: John Murray, 1885), Jeremy Bentham, *An Introduction to The Principles of Morals and Legislation.* (Oxford: Oxford University Press, 1789), Francis Hutcheson, *A System of Moral Philosophy.* 2 vols. London: privately published by his son Francis, 1775, John Stuart Mill. *Utilitarianism.* (London: Parker, Son & Bourn, 1863, rpt. 1979, Hackett Publishers); William Paley, *The Principles of Moral and Political Philosophy.* (London: R. Faulder 1785); Henry Sidgwick, *The Methods of Ethics,* 7th edition. (London: Macmillan, 1907).

II. Contemporary Sources. Annette Baier, "Doing Without Moral Theory?" in Stanley G. Clarke and Evan Simpson, eds., *Anti –Theory in Ethics and Moral Conservatism.* (Albany, NY:State University Press of New York, 1989); James Wood Bailey, *Utilitarianism, Institutions, and Justice.* (New York: Oxford, 1997); Conrad D. Johnson, *Moral Legislation: A Legal-Political Model for Indirect Consequentialist Reasoning.* (Cambridge: Cambridge University Press, 1991); Christine Korsgaard, "Two Distinctions in Goodness." *Philosophical Review*, 88.2 (April 1983): 169-95. David Lyons, *Forms and Limits of Utilitarianism.* (Oxford: Clarendon Press, 1965); _____. *Rights, Welfare, and Mill 's Moral Theory.* (Oxford: Oxford University Press, 1994); J. B. Schneewind, *Sidgwick's Ethics and Victorian Moral Philosophy.* (Oxford: Oxford University Press, 1977); Michael Slote, *Common Sense Morality and Consequentialism.* (London: Routledge, 1985); J.J.C. Smart and Bernard Williams, *Utilitarianism: For and Against.* (Cambridge:Cambridge University Press, 1973); Donald Regan, *Utilitarianism and Cooperation.* (Oxford: Clarendon Press, 1980); Judith Jarvis Thomson, "On Some ways in which a Thing can be Good." *Social Philosophy and Policy.* 9,2 (1992): 96-117.

[4] Henry Sidgwick, *The Methods of Ethics,* 7th ed. (London: Macmillan, 1907), p. 508.

[5] Of course, some like Donald Regan will demur. This is because he (and others like him) puts forth an ethos of cooperation under which utilitariansim might flourish. See: Donald Regan, *Utilitarianism and Co-operation* (Oxford: Clarendon Press, 1980).

[6] *Kraterism,* is a term I use to denote those who believe that their use of power legitimates its exercise just in case it is successful. "Might makes right" is one version of this. *Feng* is a term I use after the Anglo-Saxon verb, 'fengon' that means "to snatch or take—generally with force". The connotation is that it is a decisive and aggressive taking—rather like a "commandeering." The word is prominent in *Beowulf* as an important element of the competitive worldview of the contending parties.

[7] "Each agent must strive to create a common body of knowledge that supports the creation of a shared community worldview (that is complete, coherent, and good) through which social institutions and their resulting policies might flourish within the constraints of the essential core of commonly held values (ethics, aesthetics and religion)," from Michael Boylan and Kevin Brown, *Genetic Engineering* (Upper Saddle River, NJ: Prentice Hall, 2002), p. 16, cf. Michael Boylan, *A Just Society* (Lanham, MD: Rowman and Littlefield, 2004), chapter 5.

[8] *Basic Ethics,* p. 2.

[9] *Basic Ethics,* 19-20, 103; *Genetic Engineering,* 24.

[10] This sense of "health" goes beyond being free from accident and disease. Health is not merely the lacking of something bad, but the possession of something good (the goods of agency). This implies a sense of health as a balance. Such a notion is holistic and in keeping with the Personal Worldview Imperative. Some current work that take up issues involved in this position includes: Esther M. Steinberg, *The Balance Within: The Science Connecting Health and Emotions* (NY: W. H. Freeman, 2000); Gabriel Stux, eds. et al. *Clinical Acupuncture: Scientific Basis* (Berlin: Springer, 2001); Laurie Gairett, *The Coming Plague: Newly Emerging Diseases in a World out of Balance* (NY: Farrer, Straus, & Giroux, 1994); Andrew Weil, *Health and Healing* (Boston: Houghton Mifflin, 1988).

[11] Some would say that such a definition seems to exclude environmental concerns. Such a general stance has been taken by Carl F Cranor, "Learning from the Law to Address Uncertainty in the Precautionary Principle" *Science and Engineering Ethics* 7.3 (2001): 313-326 and Ralph Ellis and Tracienne Ravita, "Scientific Uncertainties, Environmental Policy, and Political Theory" *Philosophical Forum* 28.3 (1997): 209-231. However, I would demur to such attacks by citing my essay in which I use the above framework to defend environmental protection, see: Michael Boylan, "Worldview and the Value-Duty link to Environmental Ethics" in *Environmental Ethics,* Michael Boylan, ed. (Upper Saddle River, NJ: Prentice Hall, 2001).

[12] What I mean by an "appropriate" functional level here is really rather simple. If Jamal needs 2,000 calories a day in order to function well, then any reasonable delivery of these calories according to our latest ideas of nutrition would be functionally appropriate. Jamal need not have the very best organically grown tomatoes, but he does deserve nutritious, wholesome food that will meet his dietary needs.

[13] The caveat "as much as possible" here refers again to the "ought implies can" dictum. If there is no cure for AIDS or breast cancer, one cannot provide the agent relief from its devastations. However, this does not mean that one should be resigned to this reality. Rather, the society should devote resources to cure these diseases and work cooperatively with other societies doing the same.

[14] The "others" might include issues indirectly related to these—such as gun control is related to crime and

to domestic violence—or other categories fitting into the slots of level-one goods.

[15] This belief is difficult to assess. Some sorts of crimes are undoubtedly higher among the poor: armed robbery and murder, for example. However, other sorts of robbery—such as embezzlement and tax fraud are largely unreported and are more common among the affluent. This facet of crime is often left out of many snapshots of crime, see Steven Vago, *Law and Society* (Upper Saddle River, NJ: Prentice Hall, 2000), pp. 241ff.

[16] Racial profiling is not new and it is not confined to any particular culture. Some current discussions of this unethical behavior include: David A. Harris, *Profiles in Injustice: Why Racial Profiling Cannot Work* (NY: New Press, 2002); Kenneth Meeks, *Driving While Black: Highways, Shopping Malls, Taxicabs, Sidewalks: How to Fight Back if you are Victims of Racial Profiling* (NY: Broadway, 2000); Jacob André, *CRRF Facts about: Racism and Policing.* (Toronto: Canadian Race Relations Foundation, 2001).

[17] There is a considerable faction who advocate a wider understanding of public health. These writers advocate considerations of ethics and human rights in public health decisions. Some of these include: Ruth Macklin, "Bioethics and Public Policy in the Next Millennium: Presidential Address" *Bioethics* 15.5-6 (October 15, 2001): 374-381; Chan Ho-Mun, "Free Choice, Equity and Care: The Moral Foundations of Health Care" *Journal of Medicine and Philosophy* 24.6 (December, 1999): 624-637; John Harris, "Justice and Equal Opportunities in Health Care" *Bioethics* 13.5 (October, 1999): 392-404; Dan Brock, "Broadening the Bioethics Agenda" *Kennedy Institute of Ethics Journal* 10.1 (March, 2000): 21-38; Paul Farmer, "Pathologies of Power: Rethinking Health and Human Rights" *American Journal of Public Health* 89.10 (October, 1999): 1486-1496; James A. Morone, "Enemies of the People: The Moral Dimension to Public Health" *Journal of Health Politics: Policy and Law* 22.4 (August, 1997): 992-1020. I should also note that in a recent discussion on public health at NIH, I was confronted with the position among those who believe that epidemiology is the defining feature of public health, that using this mathematical model alone, there could be a public health of deep ocean earthquakes. Since (as per the Preface) I do not feel that epidemiology completely defines public health, this sort of "broadening" is not what I intend. Rather, it is the inclusion of morality into the field of public health just in case basic goods of agency are involved.

[18] Poverty is a condition that affects one's health in many respects. Many of these may also be linked to racism and sexism. They entail a diminishment of level-one and level-two basic goods as well as level-one secondary goods. Contemporary discussions that argue this point in more detail include, Rosemarie Tong, "Just Caring about Women's and Children's Health: Some Feminist Perspectives" *Journal of Medicine and Philosophy* 26.2 (2001): 147-162; Thomas W. Pogge, "Properties of Global Justice" *Metaphilosophy* 32. 1&2 (2001): 6-24; and William Roth, *The Assaults on Social Policy* (NY: Columbia University Press, 2002).

[19] A recent example of this in the United States is the emergence of clinics for the very wealthy. In these clinics, chauffeurs pick-up patients at their homes to ferry them to locations where there is no waiting and a physician willing to give them all the time they desire. Accommodation and convenience are the watchwords. And some in the United States say, "So what? They can pay for it." But the truth is that these salons of healthcare drain away resources from the rest of society. This is because a physician can see far fewer patients at these extraordinary rates and make more money. The rest of society's physicians would have to see more patients to make up for this loss. But there is only so much time in the day. Thus, a formula of temporal rationing occurs in which those with the least must wait longer

and longer for healthcare. All this to pamper the rich (level-three secondary goods). This is morally unjust according to the model set out above.

[20] By "developed" here I mean economically developed in the historical context of the early 21st century—such as the "G-8" countries.

[21] *Statistical Abstract,* (Washington, DC: U.S. Census Bureau, 1999), tables 144-145.

[22] The sense of "adequate" here is vague. What is intended is that fundamental health needs of all citizens should be met first. This sense of fundamental is obviously cost-sensitive. For simplicity sake I would suggest that first healthcare should be concerned with level-one basic goods. Secondly, it should concern itself with greater number of people (to be covered) over fewer just in case there are equal moral claims in which there cannot be a moral adjudication.

[23] The percentage here is relative to the effectiveness we want to achieve. The aspirational goal would be 0%. However, in the process of creating a workable policy that measures general public health sociologically, then some percentage—such as 2.5%—is useful as a benchmark.

[24] I should note here that I do NOT believe that rationality, alone, is both necessary and sufficient. The Personal Worldview Imperative sets out various understandings of the term "good." These include ethics, religion, and aesthetics (among others). The components in these various axiological criteria will certainly involve more that merely rationality. However, in the spirit of Plato's *Meno,* I would contend that all these other criteria are tethered by rationality. This means they inform and influence the scientific determination of the right and wrong in human action.

D. MICAH HESTER

PROFESSING PUBLIC HEALTH:
PRACTICING ETHICS AND ETHICS AS PRACTICE

ABSTRACT: This essay attempts to bring ethical and public health practices together in order to show that good, ethical deliberation is, by nature of the profession, good public health practice. I attempt to reconstruct some of the ethical tensions between individuals and communities by undermining the classical Enlightenment liberal notion of the self with a concept of self as social product. Such an understanding of the self does justice to the ethical character of public health practices and gives a different, more effective perspective on how to approach deliberation of public health concerns.

KEY WORDS: Character, community, habits, individuality, inquiry

In the arena of Public Health, it would seem that to ignore ethical tensions between communal and individual interests would be *prima facie* poor professional practice, for Public Health issues are constitutively concerned with the relationship between public and private "goods." Given this, it is pressing in both the practice of and education in Public Health that we avoid a common approach to most professional development—viz., *first* get the concepts and skills "right," *then* work on what is good. Instead, given the scope and character of the profession, to get the concepts and skills of Public Health professionals "right" necessarily entails working towards the "good."

However, while professional development in any field might benefit from the recognition of the embeddedness of ethics in its practices, Public Health is unique among the health care professions (at least) because, unlike all others (medicine, nursing, allied health, etc.), there is a *constant* concern for *communal* health goals that inevitably (though not universally) demands the sacrifice of individual interests. "Whereas in medicine, the patient is an individual person, in public health, the 'patient' is the whole community."[1] In other words, Public Health professionals not only necessarily consider a weighing of public and private goods, as well as an ideal of the "good life" with the appropriate means to achieve it, they do so with *community concerns as paramount*.[2]

This chapter attempts to bring ethical and Public Health practices together in order to show that good ethical deliberation is, by nature of the profession, part and parcel of good Public Health practice. In doing this, I attempt to reconstruct some of

1

M. Boylan (ed.), Public Health Policy and Ethics, 1-16.
© *2004 Kluwer Academic Publishers. Printed in the Netherlands.*

the ethical tensions traditionally accepted as fundamental to the relationship between individuals and communities by undermining the classical Enlightenment liberal notion of the self with a concept of self as social product. Such an understanding of the self, I will argue, both does justice to the ethical character of Public Health practices and gives a different, more effective perspective on how to approach deliberation of Public Health concerns.

WHAT DO WE MEAN BY PUBLIC HEALTH CONCERNS?[3]

The profession of Public Health as we know it today, grew up from a long history that (at least in Western societies) reaches as far back as ancient Greece. For example, the Hippocratic corpus[4] is replete with discussions of environmental factors in preventative care, as well as general concerns for health within a society. In general, the Hippocratic concern was for balance among bodily fluids (the "four humors" as they have come to be known) as well as between body and environment. The *Epidemics* may be the most famous Hippocratic writing in this arena, with the term 'epidemic' stemming from the Greek for something that comes "upon the people," such as the spread of disease. Subsequently, the Greek notion of viewing health environmentally, and not individually, waxed and waned in importance.

The rise of the modern Public Health profession did not occur until the nineteenth and early twentieth centuries. Discoveries and research by Edward Jenner (smallpox), Edwin Chadwick (sanitation), John Snow (cholera), Louis Pasteur (bacteria), and Joseph Lister (antisepsis) brought concerns for disease control, nutrition, sanitation, and so forth, starkly into public view. Based on a turn from a Hippocratic sense of holistic balance to a biochemical, microscopic germ model, the "state of the art" of research and medicine were changing rapidly and effectively during that time, and studies on physiological functions, disease transmission, bacterial infection, and sanitation as well as the development of useful medical instrumentation conspired to force communities to look more carefully and take more seriously the need for health-promoting conditions in everyday life.

At the same time as these research and technical revolutions were occurring, public support was motivated by moral and political concerns as much as it was by medical and environmental conditions. In particular, forefront in the development of the Public Health profession were utilitarian principles that looked towards the greatest good for the greatest number of people, a reaction motivated in part by economic industrialization and political stratification. Public Health, then, arose not just as a social need but as a moral calling.

This history, however, only begins to point to what we mean by "Public Health" as a profession and set of practices. In fact, much of the history of Public Health at such a "macro" level is simply a general account of the history of medicine writ-large. Medicine, too, has ancient Greek (Hippocratic) origins, and develops throughout the centuries with merely incidental change until the Enlightenment with its emphasis on scientific investigation and the development of new medical instrumentation. And medical history, also, culminates in professional institutionalization during the nineteenth century. Thus, one, compound question remains begged: What is Public Health and how does it differ from medical/clinical practice?

The short answer, as implied in its name and alluded to above, is that Public Health, as a discipline, looks first and foremost at *public*, not private, health issues. Focusing on groups or populations, these issues transcend particular individuals and cases. As Public Health Commissioner Lloyd Novick emphasizes, "The operative components of this definition are that public health efforts are organized and directed to communities rather than to individuals."[5] Furthermore, Public Health has taken as part of its platform an emphasis on prevention and health promotion, not treatment and cure, and these "measures save *statistical* lives and reduce *rates* of disease within populations."[6]

And yet, with Public Health's emphasis on community goals and purpose, it is not merely trivial to note that it is *impossible* for a population or community to have a health related issue that has not already affected (or has the potential to affect) some particular person. Individuals get ill, are in harms way, develop conditions, and from this, Public Health determines the communal causes or consequences. Public Health professionals also strategize about how to affect the public practices, and by that, they ultimately have to work to change individual characters.

Thus, Public Health, at its heart, must attempt to answer a central question: What is the relationship between individuals and the groups of which they are considered to be a part? Furthermore, this concern is tied up with the struggle over whose interests take precedence. Finally, Public Health works to develop appropriate means of habit creation and modification in order to promote healthy living in health environments. Given this need to adjudicate interests with a concern for communities, individuals, and environments, as we shall see below, it is no stretch to say that Public Health practices fundamentally contain ethical considerations, and thus ethics is constitutive of the Public Health profession itself.

AN UNANALYZED LIST OF PUBLIC HEALTH ISSUES

Though much more needs to be said about this, it may strike some as merely trivial or obvious to say that the profession of Public Health is, at bottom, an ethical endeavor. However, since speaking in such generalities, it may not be clear what I might mean by such a claim in the particularities of Public Health issues. Below, then, is a sample list of Public Health issues with mention of the kinds of ethical quandaries they can raise.

Issues in Public Health[7]

Communicable diseases
>   TB, STDs, AIDS, etc.

>   *Sample Ethical Issues:*
>   Privacy and confidentiality

Research
>   Disease processes, pharmaceuticals, genetics, biotech, cloning, etc.

>   *Sample Ethical Issues:*
>   "Appropriateness" of research
>   Use of vulnerable populations
>   Informed consent

Distribution of resources
>   Personnel: Nursing care, Emergency medicine
>   Technologies: Transplantations, Pharmaceuticals, ICU Instrumentation
>   Monetary: class-based access, insurance coverage

>   *Sample Ethical Issues:*
>   How to determine who gets what, when, and why?

Developing/supporting healthy environments
>   Sanitation: Environmental, Personal hygiene
>   Nutrition: Promotion, Distribution
>   Institutions: Familial, Social, Governmental

Sample Ethical Issues:
　　Weighing individual, institutional, and community interests

While hardly exhaustive, the above list shows a wide variety of issues that Public Health professionals face (and many that will be covered more precisely by other authors in this book). What is important about the list for this essay is not so much the specific issues addressed but that all these issues impress upon Public Health professionals the need to adjudicate a complex of interests expressed either at individual or communal levels. In particular, the list demonstrates that some ethical issue(s) arises in specific ways whenever a Public Health issue is addressed. Now, while this alone serves to demonstrate my general contention about the relationship between Public Health and ethics, mine is a more pervasive comment. In line with Dan Beauchamp's claim that "Public health belongs to the ethical because it is concerned not only with explaining the occurrence of illness and disease in society, but also with ameliorating them,"[8] I argue that even if no specific ethical issue can be identified through our traditional methods of labeling such things, the very nature of Public Health practices constitutes an instance of ethical deliberative practice.

Since a communal orientation is at the heart of the profession, the task of Public Health professionals can be stated simply as the need to understand and address various interests at play, while aiming towards some communal goal. As such, it might seem that individual interests always take a secondary position. Adjudications within in Public Health deliberations, then, would require sensitivity to the ethical tensions traditionally (at least since the advent of Enlightenment liberalism) understood to exist between individuals and the communities of which they are a part. However, such a take on the issue runs the risk of missing an important feature of the relationship between individuals and communities of which they are a part. And while sensitivity to individuality always is required, I believe that the traditional tensions are misunderstood because the account of individual selves that Enlightenment philosophy has left us is misguided. As I will endeavor to show below, ethical theory and ethical deliberation should arise within a framework that takes individuals, not to be in fundamental opposition with the communities, but to be socially situated products, mediated by their communities. A different account of the self as always-already socially situated changes the traditional picture and requires that our intelligent deliberations address Public Health issues (and many other ethical issues, for that matter) differently.

WHAT DO WE MEAN BY ETHICAL CONCERNS?⁹

Great authors and writings throughout the centuries (particularly in the "West") tell us that ethics and moral philosophy concern customs, principles, codes, right and wrong, good and bad, character and virtue, motives and choice, actions and consequences. It is merely a truism to note that no one account captures *the* essence of moral philosophy, and in fact, many accounts differ so greatly that outright contradiction and competition has followed. My brief account has no pretense that it is the one true account, and I do think that there is truth to be found in many different (and even differing) ethical theories.

While the English terms 'ethics' and 'morals' come from Greek and Latin roots, respectively, neither the Greek (*ethos*) nor Latin (*mores*) terms capture what we mean by our contemporary terminology. For the ancients, *ethos* and *mores* denoted "customs" or "manners"—that is, the generally accepted opinions of what are good practices within a community or culture—but contemporary ethics goes beyond mere custom and community opinion. At the same time, though, the historical connection is not difficult to see: Customs are community standards to which an individual member's actions are supposed to conform; lack of conformity results in critique, at best, and punishment, at worst. With the likes of Plato/Aristotle and Cicero, concern for *ethos* shifted to *ethike* and *mores* to *moralis* developing a broader and deeper scope.

In other words, ethics expands the *custom*ary range of concerns, placing not only individual action in question (as "custom-following" does), but the very standards of the community as well. Ethics demands a reflection upon social institutions and judgments as well as personal motives and actions. This reflection tries to adjudicate among competing interests and potential consequences, adjusting either personal motives and character or social/environmental conditions (or both) in order to determine what is (or will be) good and right. Of course, such an account of ethics begs the question, How can we tell what is "good"? And here is where much of the debate in moral philosophy arises.

To get at an answer to this, we must first look at the nature of "good." To begin, recognize that "good" functions both as a condition and an outcome, a means and an end. Some "good" drives action while "good" action is best determined from the consequences that follow. While this may sound equivocal, they are related. Motivation and character manifest themselves in our desires, and those desires are "good" to us, but what is experienced de facto as "good" is not yet what we mean by ethically "good." The moral "good" only arises as the result of reflective inquiry into the cost and consequences of pursuing our desires and interests. As John Dewey has said, that which begins as something desir*ed*—an experienced good—becomes desir*able*—a moral good—as the product of intelligent inquiry into the worth of pursuit.[10] This places inquiry—its methods and norms—centrally in field of ethics. So while ethics has, at times, been taken to be the development of a *theory* or *system* of values or principles that should guide our action towards the good, the character of ethics, as we shall pursue in this chapter, is most easily discernable through what we call "ethical conflict" and "ethical deliberation."

Admittedly, such a characterization of ethics *qua* ethical deliberation can be problematically reductionistic. Certainly, such an account runs the risk of ignoring important insights in the "groundwork" of ethics, and further seemingly marginalizes the historic concerns for "virtue" and "character" as well as the more contemporary discussions of "care." I say "runs the risk" because even a focus on ethics *qua* ethical deliberation need not ignore metaphysics or character. In fact, deliberation, taken intelligently, requires habits of reflection, suspended judgment, open-mindedness, creativity, and courage (among other things), and these habits develop and require an ethical character.

Even having said this, though, what will remain missing in my account is a *robust* discussion of what constitutes "good" and "bad"/"evil" or "right" and "wrong." However, in this context, ethical conflict requires no such account. Since competing de facto interests stimulate ethical deliberation, ethical conflict on this account is not a metaphysical battle between "good" and "evil." Instead, ethical deliberation adjudicates among competing experiential interests—or "goods"—in order to fashion what is the moral good to be pursued. Internal to the nature of ethical deliberation, then, is not the process of eschewing evil, so much as it is determining which "good" is worthy—given the costs to the environment, others involved, and ourselves—of our efforts.

We cannot better Aristotle's own account that ethical inquiry aims at some good, but what must be understood is that it is the indeterminacy of just what that good is that drives the inquiry in the first place. Ethical inquiry attempts to take this indeterminate situation of competing or unclarified goods and make it determinate. The settled outcome of such a deliberation allows us to say what is the "good" and "bad" for this situation, but, again, the deliberation itself starts in uncertainty concerning these morally charged labels.

Ethical activity is first and foremost, then, about identifying and thinking through ethical issues, dilemmas, and conflicts. That is, it is about methods of intelligent inquiry, which work to determine, at least in the situation at hand, the "right answer." It requires that we develop habits of intelligence in matters ethical, and thus, as I hinted at earlier, is concerned with the kinds of characters we create. The act of ethical character creation is, thus, deeply influenced by the kinds of education we have and our abilities to approach problems openly, with an eye towards others' interests as well as our own. One way to characterize this is that the *field* upon which ethics operates is habits of social intelligence.

Community, then, is a central moral condition and concern. To state this differently, ethics ultimately transcends mere individual interests, placing them in the social context from which they have arisen and into which they need to be considered. That is, our actions are part and parcel of our communities, and they will be judged accordingly. Meanwhile, our communities are continually shaped (and reshaped) by our actions (I shall say more on this below). In this way, we can see the deep connection between the ancient sense of *ethos* and *mores* in contemporary ethics, while we recognize further that such customs, themselves, are open to question, adjustment, even elimination in the face of a reflectively considered competing good that disrupts, if not dismantles, the rationale behind such customs.

HOW MIGHT WE (RE)CONCEIVED ETHICAL TENSIONS IN PUBLIC HEALTH PRACTICES?

As we have said above, ethical conflict demands that we adjudicate competing "goods." Ethical activity is first and foremost about identifying and thinking through ethical issues, dilemmas, and conflicts. That is, it is about methods of intelligent inquiry more so than finding the "right answer."

Also, from the outset, I stated that a concern for the relationship between individual and social interests is at the heart of Public Health practices. Mislead-ingly, however, this highlights tensions between individual and community interests, interests that Public Health issues seemingly often raise, as if those tensions are constitutive of the relationship between individuals and communities. Of course, it would be foolish to deny that such tensions exist since it is clear that Public Health goals are communally oriented, and such an orientation always runs the risk of ignoring or squelching some individual's particular desires. However, to say that Public Health operates within this relationship does not yet characterize the relationship itself.

Ever since the Enlightenment, Western thought has indeed characterized tension as part and parcel of the relationship between individuals and society. Philosophers like Hobbes, Locke, Rousseau, and Kant have championed a very useful concept of individuality that divorces individual ends from communal ones. The extent to which such ends might agree is, at most, a happy accident, since society is typically characterized as a necessary evil, forcing individuals to compromise their freedoms for the sake of a greater good or useful harmony. In this way, Enlightenment liberalism supports an ethic of atomic individuality, with social goods understood exclusively in terms of the benefit to individuals and this individualist ethic.

While useful in undermining particular kinds of politics, this characterization leaves much to be desired, resting on a dubious account of human freedom, and relying primarily on what has been called "negative liberty"—that is, liberty *not* to be disturbed or infringed upon.[11] In response, so-called communitarian approaches (by, among others, contemporary authors like Alasdair MacIntyre and Michael Sandel[12]) have been developed that champion as paramount the common good over the individual interest. Taking the alternate side of "the individual versus society" dichotomy, these writers have developed an ethic that subsumes individuality in lieu of the social factors and good that create individuals, ultimately arguing, in the most extreme moments, that individuals are nothing, save for the communal associations they represent or display.

In the face of such controversy, the issues raised in Public Health practice must be resolved in light of some view of the ethical landscape. As Sholom Glouberman has rightly pointed out, "The individual and population-based frameworks can support dramatically different policy and administrative decisions in public health."[13] And while this is true, the solution to the dilemma might seem obvious: With Public Health's unique communally-oriented view of health care, any individualistic ethic would seem to leave us wanting. Thus, the communitarian approach to ethical deliberation, where communal goods simply are the highest goods, would seem appropriate. The problem is, as I mentioned earlier, all communal goods only arise as expressions of individuals within those communities, and individual interests are part and parcel of the environment in which communal (often preventative) solutions to health care problems must operate effectively. Also neither an individualistic or communitarian approach alone can alleviate the problematic fact that Public Health has not taken seriously the social factors that influence individual behavior. As Jonathan Mann points out, "public health programs…consist of activities which assume that individuals have essentially complete control over their health-related behaviors. Traditional public health seeks [primarily] to provide individuals [only] with information and education…."[14] Thus, while Enlightenment ethical theories that focus on individuals may be obviously unacceptable in aiding Public Health professionals, it is not clear that communitarian approaches will fair any better.

In part the problem is that classical liberals and communitarians accept the same Enlightenment-generated picture of human nature with individuals and society held to be a priori at odds with one another—liberals grounding their ethic in individual interests while communitarians side with social interests as primary. However, the past hundred years of social psychology and sociology contradict the Enlightenment account of the human self as an insular, isolated being formed prior to communal relations. Instead, theories and research into human development lead us to an account of human beings and human interaction where it would be more correct to say that human selves come to be by way of relational activity with others. Rather than fundamentally opposed, then, this social psychological account uses an organic model which describes individuality as a product of social, environmental activity while still recognizing the uniquely situate and saturated selves that arise through such interactions.

To flesh this out a bit more,[15] I will turn briefly to the insights of philosopher and social psychologist George Herbert Mead who, under the influence of William James, John Dewey, and Alfred North Whitehead, determined that the self is not an entity prior to social relations but instead comes to be in and because of social processes. In a work posthumously edited from course lectures, we find an important claim about human development: "[M]inds and selves are essentially social products, products of phenomena of the social side of human experience."[16]

As Mead's psychology explains, what we call the "self," rather than being an entity upon which attributes and relations are "hung," is an organized complex of attitudes implicating both the individual and society. Certainly, biological, organic individuals are uniquely situated in and created out of complex biological processes. Rather than simply a mere biological entity, however, the *self* is a conscious, interacting being, in the world. S/he is a responsible and reflective character. The self makes distinctions and is conscious of its place in the world relative to its environment. However, these qualities do not and cannot arise until interactions with others occur. Through such interactions, the individual organism (usually in the form of a baby or young child) begins to recognize and respond to others. At first, the child simply plays games that mirror the actions of others; s/he takes on roles and characters, merely imitating what s/he sees. Children smile at our smiles, laugh because we laugh, touch what we touch. Even later, a form of this continues as they dress in our clothes, play with our tools, speak in affected voices because that is what they see and hear.

However, slowly individuals creatively separate the actions of others from their own. Rather than parroting others' actions, individuals look for responses from others to their own actions. For example, the dog's growl *signifies nothing unless* we act scared because of it. The baby's cry means that it is time to change the diaper, not because of the infant's *intent*[17], but because of the parent's (or caregiver's) *response*. The broken glass has no meaning to the child until an adult scolds him/her for breaking it; at that point, the broken glass *signifies* "trouble." Soon, the young individual becomes aware of the attitudes of others to the extent that s/he begins anticipating those attitudes in selecting gestures appropriate to the situation. Language, as an act of meaning, objectifies within the conversation the individual who is speaking; it treats him/her as an object to him/her*self.* Thus, the self first comes-to-be *reflexively.* The child says "bottle" in anticipation of the response by the parent to give the nippled object to him/her. But in saying "bottle" the child reacts to the object (if only internally) as s/he expects the parent to react. S/he leans towards it, reaches for it. The infant becomes as much a member of the audience as the parents do, listening to him/her*self.*

The self arises, then, in "self-conscious" behavior that objectifies the self to itself. This objectifying move incorporates an awareness of the attitudes of the other; that is, it takes on the attitude of the community itself. The self, then arises by way of an awareness and an internalizing of the "attitudes" of the communities of which we are a part. Thus Mead states, "In this way every gesture comes within a given social group or community to stand for a particular act or response, namely, the act or response which it calls forth explicitly in the individual who makes it."[18]

Mead's concept of "self," then, does not accept the prevailing modernist view of a prior self whose originary being comes fully formed. Instead, he takes the self to be the product of social interaction. But even this is misleading for there is no "one" self, but

> We divide ourselves up in all sorts of different selves with reference to our acquaintances. We discuss politics with one and religion with another. There are all sorts of selves answering to all sorts of different social reactions. It is the social process itself that is responsible for the appearance of the self; it is not there as a self apart from this type of experience.... There is usually an organization of the whole self with reference to the community to which we belong, and the situation in which we find ourselves.[19]

Community then is constitutive of and prior to the self. "It cannot be said that the individuals come first and the community later, for the individuals arise in the very process [of living] itself."[20] It is the taking on of community attitudes that make us "who we are" in any important sense.

Finally, this organization of the self has important moral consequences, for gestures of any significance must recognize and respond to others as we take on their attitudes as our own. And the meaning of our actions come, not by way of our intentions (though they may arise from our own impulses) but in how they are taken by others— that is, how they bear out in their consequences.

If we look now towards the end of the action rather than toward the impulse itself, we find that those ends are good which lead to the realization of the self as a social being. *Our morality gathers about our social conduct. It is as social beings that we are moral beings.* On the one side stands the society which makes the self possible, and on the other side stands the self that makes a highly organized society possible. The two answer to each other in moral conduct.[21]

Moral activity occurs among social beings aware of this social self. Moral conduct and judgments must themselves be social such that "one can never [judge] simply from his own point of view. *We have to look at it from the point of view of a social situation....* The only rule that an ethics can present is that an individual should rationally [and imaginatively] deal with all the values that are found in a specific problem."[22]

This social character to ethical deliberation has already been mentioned earlier, and we can now see how such a position follows, not from a need to avoid interfering with others, but specifically because we are so deeply entangled with them.

However, it is important to note that, given the vast number of interests and values at play in the world as well as the limited and finite character of our abilities and the universe itself, tragedy is inherent in ethics. From the infinite number of interests at play in the world and the finite time, space, and resources available to us follows that there is always a risk that someone's interests will be destroyed whenever we act through our morally deliberative outcomes. From the perspective of those whose interests are left behind by deliberation, much is lost. So much the worse for them, and as we relate to them and are deeply connected to others of our community, so much the worse for us as well. Avoiding such loss whenever possible, then, is a goal devoutly to be wished. That is why ethical deliberation demands a creative spirit. As William James has put it, "*Invent some manner* of realizing your own ideals which will also satisfy the alien demands—that and only that is the path to peace."[23]

Since my activities are never exclusively my own—i.e., they arise, in part, from the social conditions in which I find myself and will consequently affect others of my social group—if I wish to perform my actions "to the good," I must account for the many (and often competing) interests at play in the situation. Those interests arise from other selves who are part of the environment in which I wish to exercise my own (communally constituted) desires.

Luckily, while this is a subtle, even difficult, position to understand, aid to Public Health professionals in practical adjudications is completely in line with the insights above and are already available through and to Public Health itself. We have noted already that Public Health professionals have long understood the intimate connection, generative nature, and constitutive relationship between communities and individuals, but such insights must form habits of communal individuality in decision-making. It would seem that the danger for Public Health professionals should rarely be that they would forget about community interests in their deliberations, but that individuality would be ignored completely. Respect for individuality is important, and must be maintained by any activities and policies Public Health develops. However, any ethic that makes individual interests paramount would seem to undo the entire purpose of Public Health. While community interests should not entirely subsume individual ones, individual interests are the result of and themselves produce a complex of social factors and environmental processes. Rather than accepting a fundamental dichotomy between individuals and society, then, a view of individual selves as communal and socially situated privileges neither individual nor communal interests a priori. In fact, on this account, respect for individuality demands the development of a healthy environment that enables respectful activities since to be an individual is to be a particular socially located self, and to respect such an individual is to facilitate and enhance those specific social (and reflectively acceptable) interactions that constitute that self.

Immediately, implications for Public Health professionals arise. Since Public Health professionals deal with difficult social issues from a unique perspective, and since their decisions must take into account complex societal contexts and many group and individual interests, it should not be surprising that at least some time should be spent developing an ethic for Public Health practices and policy that is sensitive to those practices and the environments in which they (attempt to) operate. Furthermore, like all ethical dilemmas, "Choices in health policy are often between one 'good' and another."[24] "Inventing" ways of adjudicating these competing "goods" by working from the perspective of communally situated individuals must be at the heart of professional activity. That is, Public Health professionals should not operate as if a fundamental dichotomy between individuals and society exists, for if they do, their task has failed before it begins. If individuals *necessarily* loose when communal interests are followed, then why should individuals ever support public policies? On the other hand, taking the perspective of an ethic of socially situated beings, forces us not only to take seriously the communal character of individuals and their interests, at the same time we must stress the important effects on individuals in the community. Though Public Health is communally focused, individuals are deeply affected by any communally operative decisions, and decisions that not only take into account the interests of individuals and groups, but try to fashion a solution that speaks to all interests at play, fitting them together into an organic/environmental whole, are decisions best able to make for long term, healthy solutions.

Questions like, "How does a deeper understanding of an individual's communal orientation, from the perspective of that particular individual, affect her care or public policy that addresses her care?" must be asked. Our surveys and instruments of measure in Public Health must allow for deeper and broader, more varied and complex understandings of people and relationships in order to do justice to community needs and the unique individualities that those communities produce.[25] The more explicit about why such groupings are being made, and the specific characteristics focused on, will make the adjudication of not only research but ethics better.

Finally, it is important to note that the recognition of the constitutive relationship between individuals and society, as well as the need to work with all interests involved (noting the depth of those interests, adjudicating them respectfully) is as much attitudinal as it is productive. That is, as Aristotle knew long before our time, habits and character-development matter. Public Health as a profession operates in the world around it, and the character of those who practice affects the character of the profession as a whole. We shall know them by their fruits, but those fruits will arise most easily and most robustly in fertile, rich soil.

My discussion herein, then, has not been primarily about issues specific to Public Health professionals as it is a description of the nature of ethics and Public Health for the sake of pointing a way to habits of ethics. To the degree that I have been successful in relating Public Health practice and ethical deliberation, I hope to have succeeded in demonstrating that habits of ethics are habits necessary to the practice of Public Health professionals as well.

NOTES

[1] Dan E. Beauchamp and Bonnie Stienbock, *New Ethics for the Public's Health*. (New York: Oxford University Press, 1999), 25. Cf. Jonathan M. Mann, "Medicine and Public Health, Ethics and Human Rights" in Beauchamp and Steinbock 1999, 83-85.

[2] Cf. Bernard J. Turnock. *Public Health: What It Is and How It Works*. (Gaithersburg, MD: Aspen, 1997), 14-21.

[3] This brief account is not intended as a thorough history of or introduction to Public Health practice, but I wish to show that Public Health practice, while containing concerns and issues unique among the health professions, is intimately connected with ethics. For a robust but not-too-lengthy history of Public Health see Theodore H. Tulchinsky and Elena A. Varavikova, *The New Public Health: An Introduction for the 21st Centruy*. (San Diego: Academic Press, 2000), chapter 1.

[4] While often attributed to the character Hippocrates, the many volumes of writings under that name are surely from a variety of authors. Cf. Michael Boylan, "Hippocrates" in *The Internet Encyclopedia of*

Philosophy. http://www.utm.edu/research/iep/h/hippocra.htm.

[5] Lloyd F. Novick, "Defining Public Health: Historical and Contemporary Developments" in *Public Health Administration: Principles for Population-Based Management*. Novick and Mays (eds.). (Gaithersburg, MD: Aspen, 2001), 3.

[6] Beauchamp and Steinbock 1999, 25.

[7] For a different kind of list of ethical issues in Public Health cf. Tulchinsky and Varavikova, 2000, 767.

[8] Dan Beauchamp, "Community: The Neglected Tradition of Public Health" in Beauchamp and Steinbock 1999, 65-66.

[9] Even though I do not intend this sketch of ethics, ethical concern, and ethical deliberation as a primer in ethics and ethical theory, it is intended to provide enough to lay some ground for connections to be made later.

[10] Cf. John Dewey, *The Quest for Certainty* in *The Later Works, 1925-1953*, vol. 4. Jo Ann Boydston (ed.). (Carbondale, IL: Southern Illinois University Press, 1988) Chapter 10.

[11] Cf. the classic statement by John Stuart Mill, *On Liberty*. Elizabeth Rapaport (ed.). (Indianapolis: Hackett Publ., 1978), chapter IV.

[12] Cf. Alasdair MacIntyre, *After Virtue*, 2nd ed. (Notre Dame, IN: University of Notre Dame Press, 1984) and Michael Sandel, *Democracy's Discontents: America in Search of a Public Philosophy*. (Cambridge, MA: Harvard University Press, 1996).

[13] Sholom Glouberman, "Ethics and Public Health" in *Public Health Administration: Principles for Population-Based Management*. Novick and Mays (eds.). (Gaithersburg, MD: Aspen, 2001), 156.

[14] Mann 1999, 86 (see fn. 1).

[15] The following paragraphs on Mead are edited down from an earlier work. Cf. DM Hester, *Community As Healing: Pragmatist Ethics in Medical Encounters*. (Lanham, MD: Rowman & Littlefield, 2001), chapter 4.

[16] George Herbert Mead, *Mind, Self, & Society: From the Standpoint of a Social Behaviorist*. Charles W. Morris (ed.). (Chicago: University of Chicago Press, 1962 [1934]), 1.

[17] We could say that children cry when they are "uncomfortable," but the *cognitive* character that even this minimal description implies is simply not there for most infants most of the time. Children cry, and they *know* not why, but they do in fact *have* experiences that result in "discomfort" and crying.

[18] Mead 1962, 47.

[19] Mead 1962, 142-143.

[20] Mead 1962, 189.

[21] Mead 1962, 386 [emphasis mine].

[22] Mead 1962, 387-388 [emphasis mine].

[23] William James, "The Moral Philosopher and the Moral Life" in *The Writings of William James: A Comprehensive Edition*. John J. McDermott (ed.). (Chicago: University of Chicago Press, 1977), 623.

[24] Tulchinsky and Varavikova, 2000, 769.

[25] For example, something as basic to Public Health as the "identification" of "groups" that are affected by diseases, recourses, and so forth, immediately raises ethical concern that such an "identification" has an effect on individuals who are so categorized. What if those individuals do not identify with the groups in which they are placed? Does the very act of grouping individuals cause marginalizaiton, stigmatization, discrimination? While it would be impossible to do effective Public Health work without such "grouping," it is vital to note the value-laden nature of such a practice and remain sensitive and adaptive. Any particular individual may not agree with communally developed characterizations or goals; however, every individual is a socially situated product, a result of communal forces and attitudes that uniquely arise in and through biological and psychological processes.

EDMUND D. PELLEGRINO *and* DAVID C. THOMASMA

THE GOOD OF PATIENTS AND THE GOOD OF SOCIETY: STRIKING A MORAL BALANCE

ABSTRACT: The relationship between the good of individual patients and the special good is examined when they are in conflict. The proposition is advanced that the ethical resolution of such conflicts requires an ethic of social medicine comparable to the existing ethic of clinical medicine. Comparing and contrasting the obligations clinicians incur under both aspects of the ethics of medicine is propadeutic to any ordering of priorities between them. The suggested partition of obligations between patient good and the common good is applicable beyond medicine to the other health professions.

KEY WORDS: clinical ethics, social ethics, patient good, social good, ordering priorities

I. INTRODUCTION[1]

In previous works we have held that an authentic ethic of clinical medicine must have its roots in a philosophy of medicine in which the good of the patient determines the obligations and virtues of the health professional.[2][3][4] In this essay we extend the same line of reasoning to the medicine of society. We contend that an authentic ethic of social medicine must have its roots in a philosophy of society in which the common good determines the obligations and virtues of the health professional. We deem a parallel development of the ethics of individual and social medical ethics to be a requisite for any ordering of priorities between, and among, them when they come into conflict in decision making.

Though the ethics of medicine has traditionally centered on the obligations of physicians to individual patients, there has always been a need to recognize the ethical issues arising from the fact that medicine is always practiced within a social context. The factual basis for the recognition of this fact was late in coming in the history of medicine.[5] It is, however, especially pressing today for several reasons.

Physicians and nurses today practice within organizations, institutions, and systems; they are members of interprofessional health care teams and professional associations; access, availability, and distribution of health care has become a question of justice, and fairness; the economic, societal, and political impact of medical decisions have ethical significance, as does the conduct of health care organizations; potential

M. Boylan (ed.), Public Health Policy and Ethics, 17-37.

and actual conflicts between the good of individual patients and the good of society are realities in managed care and in proposed systems of health care reform. As a result of all this, some bioethicists even suggest that the focus of the physician's ethics should be society, not the individual patient.

One of the more important questions raised by this recent recognition of the social context of medicine is the resolution of conflicts that may arise between the individual patient and the social common good. Can physicians serve the good of both? Can clinicians preserve their dedication to the primacy of their patient's good and still take the common good into account? Is some ordering of priorities possible when two good ends come into conflict? Is there some way to achieve moral balance between ends? Are those bioethicists who urge displacement of the physician's obligations from his own patient to the good of society, or entrust the good of the patient to institutions rather than physicians, to be heeded?

Responding to such questions requires a philosophy and ethic of social medicine comparable in development to that of clinical medicine. This essay approaches some of these questions in the following way: first, we examine the nature of social medicine and its dependence on a philosophy of society; then we digress to discuss the philosophy of ends, since our approach to the answers we suggest is teleological; we then move to an analysis of the composite nature of the good of society as the end of social medicine with a reemphasis on the distinctions between the functions of the clinician and public health physician; we close with a suggested ordering of the physician's obligations both to patients and society when these are in conflict.

II. THE MEDICINE OF SOCIETY WITHIN A PHILOSOPHY OF SOCIETY

The term "social medicine" has had a troublous history. It has been equated with the diseases of civilization, an anti-modernist ideology, and the study of the effects of lifestyles, culture, and environment on health and disease. As Porter points out, these themes are often intermingled.[6] For this reason we prefer the term the "medicine of society" for what others might today subsume under "social medicine" but will use them interchangeably.

By the medicine of society we mean simply the use of medical knowledge to the advance the health of society, of humans living collectively in households, families, communities, and nations. This is distinct from clinical medicine, which is the use of medical knowledge to help, heal, and relieve the sufferings of individual patients. We shall use "social medicine" and the "medicine of society" interchangeably in this limited sense without denying the importance of the many other dimensions that might be included under the same rubric.

The ethics of the medicine of society, to be properly delineated, should be located within a broader context of a philosophy of society. We prefer this term to a social philosophy, which is currently used too diffusely for our purposes.[7]

By a philosophy of society we mean a study of the nature, being, and existence of humans living and working together. It is studies of the organisms humans generate to fulfill their essential nature as social and political beings, beings who need society and social instruments to attain their good as humans. The locus of study of a philosophy of society may be the family, community, state, nation, profession, or even the global community. A philosophy of society begins with the question—"What is society, what is its nature, to what does it tend, and what is its telos or end?" The telos of society is ultimately the good of the persons who constitute that society, the good essential to their fulfillment of their potential as humans. This is a good that cannot be fully achieved by humans living isolated from each other.

Within such a philosophy of medicine the medicine of society has a specific function. That function is the use of medical knowledge to cultivate the health of the social organism by treating illness and preventing disease in its members since a healthy society cannot thrive without healthy citizens. An ethic of the medicine of society is directed to the good of the social organism, to the common good, the good shared by all and owed to all.

To be sure, the ethic of the medicine of society will be shaped by the philosophy of society within which it exists. In a libertarian society conceived as a voluntary association of free individuals (gesellschaft), the ethics of social medicine will be constructed in terms of free markets, individual choice, and little or no government involvement. In a communitarian society (gemeinschaft) in which the individual is defined by the group, the ethic would emphasize just distribution of goods, controlled markets, limitations on individual freedoms, and government involvement. In each case, the well functioning of society and its members is sought.

The philosophy of society that provides the framework for the ethic of the medicine of society that we espouse lies between these extremes. It is rooted in the social philosophies of Aristotle and Thomas Aquinas. This social philosophy holds to a reciprocal view of the relations of the good society and the good person. Neither has sovereignty over the other. It avoids totalitarianism, which exalts the common good above the individual as it avoids anarchism of exalting the good of the individual over the good of the whole. A truly dynamic philosophy of society recognizes the necessity of a continuously negotiated struggle to balance individual and common good.

Within this dynamic relationship of individual and common good, health and health care can be seen as societal goods because health is a good of human life, an essential component of human flourishing. In his *Politics* Aristotle speaks of the special care that should be taken of the inhabitants of a society.[8] In establishing a city he lists health as a first necessity.[9] His reference here is not just to providing individual care but to the public health as a common good.

Aquinas, likewise, takes the function of the State to be the promotion of the common good which he specifies in terms of preservation of peace, promotion of moral well being, and ensuring a sufficient supply of the material necessities of life.[10] According to Aquinas the State, like society, is necessary for the development of human potentialities and its function is to provide the conditions for the good life.

Clearly, the conception of society set out by Aristotle and Aquinas is incompatible with the extremes of a libertarian, laissez faire conception of society or, on the other hand, with a Marxist, all-consuming state-controlled economy. For both Aquinas and Aristotle the good for humans and the good for society are not determined by social preference. Rather, the good is defined by natural law that sees societies and life in communities as essential for humans if humans are to develop their full potentialities as human beings.[11]

Neither Aristotle nor Aquinas could imagine the enormous capabilities of today's medicine, which when properly used, can enhance social and individual flourishing. But it is not unreasonable to assume that they would regard health and medical care as among the responsibilities of a good society toward its citizens, but not their highest good. Health would be at the least a material and instrumental good for both the individual and the society. At best it would be a material necessity that the State should assure for all. Health care could not be a privilege to be enjoyed only by those fortunate enough to afford it. It could not be left to the fortuitous interplay of commerce, the competitive marketplace, and the medical entrepreneur.

In a good society health care is a common good as well as an individual good. Herein lies the tension that is of such growing concern today when health care resources are generally regarded as limited relative to the potential benefit they offer if used optimally. That tension brings commutative and distributive justice into conflict.[12] Traditionally the physician has felt ethically bound to commutative justice, i.e. the obligation to be faithful to a promise of trust that he or she will act primarily in his or her patients' best interests. But, in recent years, increasing pressure from governments, health plan administrators, ethicists, and the public have tended to add distributive justice, i.e. the preservation and conservation of social resources to the physician's ethical obligation.

Some ethicists and policy makers suggest that a "new" medical ethic is necessary, one in which the physician's ethical concern should be transferred from the primacy of the patient to the primacy of the society.[13] A further extension of this trend is to move the patient's trust relationship away from doctor to the institution. The health "system," not the physician, in effect becomes the patient's healer, advocate, and guarantor of safety.[14]

Our line of argument rejects these calls for a "new" ethic of medicine. It also resists trends to establish societal duties as primary for clinicians. We acknowledge that medicine as a practice, and physicians and health professionals within that practice, do have social obligations. Nonetheless, these obligations can, and must, be served without sacrifice of the trust relationship inherent in the clinical encounter. We therefore distinguish the obligations of the clinician that are dictated by the ends of clinical medicine and those of the public health physician or nurse dictated by the end of the common good.

The clinical relationship centers on a vulnerable, anxious, dependent, often suffering individual person. By offering to help, the clinician "professes" to possess medical knowledge that she will use for the patient's good. The clinician serves the common good by her dedication to the good of individual patients. Clinicians, physicians, and nurses are de facto advocates for the good of their patients.

For public health physicians and nurses the relationship is with the whole society. The end or purpose of the relationship is the good of humans as a collectivity, the common good. Public health physicians act for the good of all to the extent that medical knowledge can serve that good. They are the de facto advocates for the common good. Their "patient" is society and its ills. They serve the good of society's individual members secondarily by assuring a healthy community in which the individual can flourish.

Clinical medicine and public health medicine having different immediate ends cannot be conflated. They remain in a dynamic relationship with each other since the end of each is essential for human well-being. This is consistent with the social philosophy we have espoused above. Clinical medicine and the medicine of society, however, can in exercising their obligations, each within its own domain, conflict with each other.

That conflict may be generated on either side of the relationship. In the one case the undeviating commitment of the clinician on the good of his patient can conflict with societal attempts to conserve resources, impose standards of clinical care, or provide tort relief for medical error. By the same token, the efforts of those who practice the medicine of society may conflict with the pursuit of patient good by over-regulation of bedside decisions, limiting hospital access, or providing inadequate mental health care for the poor, or overburdening clinicians with paperwork that takes time from care of patients. On the social philosophy we have espoused practitioners of clinical medicine and of the medicine of society both serve a human good, each from its own perspective. When they do conflict in fact, there is need for some ethical priority setting.

Such a setting of priorities requires a framework in which the ethical foundation for both clinical and social medicine can be interrelated. Much of the history and literature of ethics and bioethics consists of elaborations of the ethics of clinical medicine and individual patient care. Similar frameworks for the ethics of the medicine of society are still in a state of development. In the next section of this essay we offer a philosophy and ethic of the medicine of society based in a definition of the ends of social medicine.

First, a word about ends is necessary because we ground individual and social ethics of medicine in the ends that distinguish them.

III. A WORD ABOUT ENDS

Today's confusion about the ends of medicine and the need for their redefinition lies in the erosion of the Classical-Medieval notion of ends, their relation to the good, and the relation between the idea of the good and ethics.[15] The good is the end or telos of human activity, and the end is that for which a thing exists, that which an act is designed to bring about. Ends are rooted in the nature of things themselves. They answer the question "What for?"[16] We do not impute ends to things; things are not good because we desire them. We desire them because they are good. We may put things, like medicine, to certain goals and purposes, but whether these are good or bad uses depends upon whether they fulfill the ends for which medicine exists and that define it qua medicine.

Aristotle and Aquinas, whose line of reasoning we follow here, were concerned chiefly with the larger conception of the good for humans as the end of human activity. Both structured their moral philosophies on the good as the end of human life. That end in its ultimate sense was, for Aristotle, a life consistent with the natural virtues, which led to happiness. For Aquinas, it was a life lived in accord with the natural and spiritual virtues that led to the beatific vision and fulfilment of the spiritual nature of humans.

Both Aristotle and Aquinas used medicine as an example of a human activity with a definable end and good, a lesser good, of course, than the ultimate good of human beings as such. They defined the final end of medicine as health, toward which the activity of medicine tended, that which made it what it was, and distinguished it from other human activities. Yet health was for them a subsidiary end, oriented toward the life of an individual in society an enhancing as many of that individual's powers of fulfillment as possible.

Thus, in determining the ends and good of human life, and in the realm of lesser good in everyday life, ends and the good are intimately related. Today, discussion of ends has been replaced by discussion of values and choices. The rights to choose and to value have become the warp and woof of bioethics, rather than a search for the good of individuals and society. Iris Murdoch put it this way: "The philosopher is no longer to speak of something real and transcendent but to analyze the familiar activity of endowing things with value."[17]

The shift from consideration of ends to consideration of "value" choices lies at the root of confusion about social medicine and its philosophy as well. On the modernist view, social medicine should be aimed at whatever people value or choose among the sentiments of liberal society. The continuing debate about prescribing growth hormones for healthy, but smaller than average children, is an example of how social mores about size and its importance directly influence clinical medicine and public policy. The debate about the proper use of this and other capabilities of modern medicine, like so many others, will be interminable if it is not anchored somehow in a notion of the good for humans as it relates to the powers of modern biotechnology. The ongoing debates about "enhancement" versus "treatment" are an example of our society's confusion about the proper ends and uses of medical knowledge.

IV. THE GOOD OF SOCIETY—A QUADRIPARTITE END OF MEDICINE

In one of our books we defined four levels in the complex notion of the good of the patient in the clinical encounter.[18] In an analagous way we can develop a quadripartite notion of the ends and good of social medicine: 1) The first and lowest level is the medical good of society, that can result from the application of medical knowledge to cultivate the health of society as an organism; 2) The second level is the good of society as society perceives it; 3) The third level is the ontological good of society qua society; and 4) The last level is the spiritual and non-historical good, that which fosters the flourishing of the human spirit. Taken together, these four levels of social good anchor the ends of medical knowledge when it is applied in a social context.

The medical good of society relates most closely to the techné of medicine, nursing, dentistry, etc. It is the good determined as indicated by the current state of medical knowledge, by what is subsumed under the rubric of the standard of care. The good of social medicine is aimed at the medical good of the social organism as a whole: Prevention of disease and disability, assuring a healthy environment, containing and ending epidemics, public education in matters of health, advising appropriate agencies on such matters as occupation health, safety of food and water supplies, occupational health and safety, responding to natural or man-made catastrophe, etc. In short, all those domains subsumed under the title of public health and social medicine are dedicated to the medical good of the social organism.

On this view, then, the medical goods of society differ from the medical goods of clinical medicine only in scope, not in kind. The training of the health professionals at this level focuses especially upon dealing with the larger forces operating in groups and communities. The associated moral problems of medicine at this level center on the difficulties involved in adjudicating the proper balance between providing these goods for the sake of the entire community, based upon its needs, and the other levels of social goods and services beyond health, such as education, housing, etc. Even in nations that provide access to health care for all, elements of distributive justice must be considered so that the health budget does not compromise the resources available for other social goods not related to medicine.

In clinical medicine the medical good can actually become harmful. If it is provided solely on the grounds of clinical or physiological effectiveness, it may result in harm, overtreatment, etc. So too, the medical good or society cannot be allowed to overmaster other social goods that may matter more to communities. A good example was provided in the public discussions leading to the Oregon "experiment," when senior citizens covered by Medicaid chose to put resources into home visits by health professionals to check medications rather than into access to emergency room care. This was their perception of their good (our second level). Yet, were they to experience a medical emergency, like a stroke, heart attack, hemorrhage, etc., emergency care might well be their first choice.

In this case, the good defined by purely medical criteria conflicted with society's perception of its medical good. It is arguable whether or not this was socially the best choice. Yet in the allocation of resources, the final decision rests with society and not the physician. We would argue that in this case the medical good of society was compromised. Robert Veatch, however, might argue that whatever society decides is a good, should be provided by health professionals. Social consensus, he contends, determines the good, not the physician.[19]

We think this is an error of delegation. Health professionals are trained to determine the medical good of individuals and society. Society may reject their choice but this it does at its own peril, just as the patient does who rejects effective antibiotic treatment from an infection. The task of the health professional is to provide information necessary for a rational policy choice. Society's perception of its own good may differ for many reasons, especially in the allocation of its resources. While the final decision is a social one, the health professional must retain a personal and professional integrity as a critic of the scientific and technical content of that decision.

The Good as Perceived by Society—Examples of a Conflict

Society is, thus, not the final arbiter of the medical (scientifically indicated) good of society, just as a Jehovah's Witness patient does not determine the clinical medical good of receiving a blood transfusion. She deems the medical good to be a spiritual harm. However, the religious patient accepts or rejects the scientifically-based medical good for the sake of her perception of the good at a higher level. Similarly, society may balance provisions of social medical good based upon higher values.[20] For example, in the current world epidemic of Corona virus infection (SARS), rights of privacy are seen by some as endangered by certain quarantine regulations.

Thus, like individual patients, society may not perceive the medical good, as defined by health professionals, as "good." "Society" may prefer other good things it perceives as preferable to the health care – economic growth, the ability to compete economically with other nations, military service, public safety, liberty in personal choice and risk taking, education, housing, recreation, etc. Seat belts, car seats for infants, safe driving, restrictions on hand guns, abstinence from tobacco and addictive substances, etc., have been widely promoted by the medical profession as good for the social organism as a whole. They have often been neglected in favor of freedom of choice, economics, or lifestyle preferences.

We have just mentioned and given examples of how patients and societies may perceive other goods to be more important than the medical good being suggested or recommended to them, and how this dynamic is also part of the social dynamic in health care. The social medical good serves the many complex facets of what individual societies may perceive as their own good. At this level a social philosophy of medicine would be concerned with political choices, preferences, and concerns that may distinguish one society from another. Here society determines the balances it wishes to provide its citizens, and its political processes should facilitate to public dialogue and decisions about that balance.

Although medicine is a universal discipline and is practiced world-wide, it is not the duty of the physician to make these social choices except as he or she functions as a citizen, an invited consultant, or as an agent of the government in carrying out its social policies about health care. Since each country and society is unique in its demographics and customs, they will balance the social medical good in different ways. These choices are determined by interactions between and among citizens, their specific economic and natural environments, and their cultural and religious histories. Just as individual patients might decide how a specific treatment fits into her life-plans, so too, society decides what elements of health care fit into its own plans for human development.

The Good of Society qua Society

All social good must ultimately be related to the general good for human societies. Humans are social animals and need a healthy society to sustain their flourishing; similarly, a healthy society is not possible without healthy citizens. Clinical medicine and social medicine intersect in preserving the dignity of the human person. To do so, each must respect human rationality and freedom in decision-making. To serve the good for society as society, health professionals must foster the inherent value of the person independent of wealth, prestige, education, and social position. In clinical medicine the patient is a fellow human being alongside the health professional. They are joined together at this level by bonds of solidarity, trust, and mutual respect. At the societal level, doctor and patient are united by the same bonds with the whole of society and ultimately with all humanity.

It is at this third level that many of the familiar principles and concerns of biomedical ethics are philosophically rooted, such as respect for autonomy, beneficence, non-maleficence, and justice. These are principles which a society, concerned with preservation of the dignity of its citizens as humans, must assure. At this level, the qualities of justice are to be observed with respect to health care. Justice here is understood as the equality of treatment of all human beings who are equally vulnerable with respect to illness and death. One of the axioms of moral medicine is that each individual person must be treated as a class instance of the human race. This axiom is applicable to both clinical and social medicine.

Denial of care to the poor or disvaluing the lives of handicapped persons, for example, violates their inherent dignity as human persons, not just their "share" of the health care marketplace. Intentionally putting some members of society at risk presumably to help others, without their consent, as was done in radiation experiments in our country, is a violation of fundamental human rights.[21] The newly-developing efforts in bioethics to reintroduce global, environmental, and international human rights concerns would also be placed at this level.[22]

Thus, the first and second level, i.e. social medical good and perceived goods, must be related to the third, the good for human beings as human beings. At this level, both clinical medicine and social medicine intersect in the good end of preserving the dignity of the human person, by respecting his and her rationality and decision-making. They recognize, especially, the inherent value of the person as independent of wealth, prestige, education, social position, and other characteristics that so often serve to separate rather than unite human beings. As in clinical medicine, the patient is a fellow human being with the health professional. They are bound together at this third level of good by bonds of solidarity and mutual respect.

In a more classical sense, the prima facie principles of contemporary bioethics and the universal rights of humans, as enunciated by the United Nations, come together in the natural law. The good of society and the good of each person in that society are mutually re-enforcing. They link the good of man (and woman) with the good of society in a dynamic tension. They underlie the characters of the good society and the good person.

At this level, justice requires that health care be treated as an obligation of a good society – as a moral obligation of a good society to its citizens. This is because health care is a universal human need – a need all humans experience if they are to lead fulfilling lives and be cared for when they are ill. Each citizen, thus, has a claim on his fellows – not to health, but to care when he or she is ill. Health care is in essence an obligation a good society owes its citizens in justice.

Spiritual Good of Society

The fourth and highest level of good for clinical encounters, between patient and health professional, is the spiritual good of the patient, as we have noted.[23] This is the good of the patient as a spiritual being who transcends ordinary material concerns. Analogously, there is a spiritual dimension in the community itself, though it is more difficult to define. This dimension is always present, but it becomes more visible in times of crisis, for example, after a terrorist attack like that of September 11, 2001. For some thinkers, the social ethic of medicine stands or falls on the adequacy of its articulation with deeply embedded spiritual values inherent in the very notion of the community.[24 25] For Loewy, the spiritual dimension is compassion that must emerge from the fact of suffering of all creatures;[26] for Welie, it is the intersubjectivity of suffering and shared values that grounds the clinical encounter;[27] for Jensen, it is the brotherhood of a common culture and concern;[28] and for still others, it is the solidarity with the sick and the potential for human development. For Christians, generally it is the solidarity of all humans as children of the same Creator.

The spiritual good of a society encompasses the transcendent principles of the culture. It gives ultimate meaning to human lives. It is that for which humans will make the greatest sacrifices of other good things to preserve. From the perspective of the structures of human existence, the spiritual destiny of the human being is the highest and ultimate good.

For many cultures this will mean the religious beliefs of their citizens. For example, despite the physical need to examine a pregnant woman in clinical medicine, a different method of examination might be required in some Islamic societies where privacy is a religious value. Or, even though a respirator may be appropriate for a patient suffering from severe trauma, for the Navajo American Indian, this may violate a profound religious belief about God and nature and be proscribed.[29] This is not to assert that all cultural practices should be tolerated simply because there exists a religious tradition to support them.[30] At the very least, however, efforts must be made to understand and if possible accommodate the lower order of medical good to the higher order of the spiritual good of individuals and societies.

V. METHODOLOGICAL REFLECTIONS

Thus far we have argued that the public aspects of medicine, and the social ethics of medicine that results, may be interpreted within the same framework as that of the clinical encounter. By keeping the anchor in the clinical encounter as we have in this chapter, we tried to avoid the contemporary tendency to over-medicalize all of society's problems. Physicians need not be, indeed, should not be social engineers, as the Nazi experience so clearly taught us.[31]

For example, domestic violence contributes enormously to emergency room admissions. Clinicians have a duty to address this violence within the realm of their expertise and the clinical case. Nonetheless, not all physicians need take on the public health features of this violence. Otherwise their time would be consumed and their other responsibilities to those in immediate need, would be neglected. Therefore, there is need for clinicians to observe a certain economy of pretension with respect to the frequent and obvious social dimension of their practice.

For us this "clinical parsimony," means that a social ethic of medicine might address itself to the social issues encountered clinically. But the "patient" is now society, and the good is the public's health. The duties and obligations, the characteristics and virtues, of public health physicians are, if not the same as those that engage individual patients are at least analogous. This is not to deny that many illnesses experienced by individuals are caused by social problems, such as poverty, ignorance, poor hygiene, lack of access to safe water, and the like.[32] Indeed, these broader causes of illness and disease quite rightly are the subject matter of public health and social medicine, and reflect the adequacy of the provision of health care in any society. Thus, in the model of a social ethic of health care we are using the causes and effects of health and disease include social and even cultural conditions.

Other methodologies are clearly possible in deriving a philosophy of social medicine. Some are based upon the idea of limiting clinical pretensions that has guided our thinking. But there are other methodologies that tilt the balance between individual and social concerns in favor of the latter. For many European thinkers, the focus of a philosophy of social medicine is less on its analogies with clinical medicine, and more on the power it engenders in its public relationship to both society and individual patients.[33] The starting point of these models of a philosophy of social medicine is social power, its dominance and frequent arrogance, and the need to reign it in.[34] Foucault's empirical philosophy of the clinic fits this model. It is a heuristic theory that not only describes current practices, but also relates these in theory to one's own experiences of illness and repair.[35]

Feminist bioethics might also be seen as another form of this view, insofar as it focuses upon the power relation of gender within medicine and the vulnerabilities that arise from differences in the social status of genders. Similarly, philosophies of medicine that concern the rise and power of technology would be examples of this kind of philosophy of social medicine.[36]

VI. CLINICAL AND SOCIAL MEDICINE: SOME FURTHER DISTINCTIONS

In commenting on Aristotle's *On Sense and the Sensed Object*,[37] Aquinas grapples with the distinction between the particular and individual in medicine and the more universal causes and effects of medicine. He notes:

> ...it is for the physician to consider their [universal principles of health and disease] particular principles; he is the artisan who causes health and like any art his must concern itself with the singulars that come under this project, since operations bear on singulars.[38]

Clinically-oriented physicians are primarily oriented to particularities. Nonetheless, the "universal principles of health and disease" are discoverable, analyzable, and manipulatable by persons other than those professing to heal individuals. Basic scientists, for example, are charged with, or have taken on, the laudable goals of improving the health of all human beings. Furthermore, examining the structures of illness and healing are at least in part the goals of a philosophy of medicine. The individualities pursued by physicians arise from more general causes and return back, mutatis mutandis, to existential structures of human existence through the individual.

A good philosophy of social medicine, then, will not neglect the centrality of the clinical encounter as the origin of questions about general causes of illness and disease, as well as the social effects of their neglect or their alteration.

In this regard, Aquinas notes further:

Health can only be found in living things, from which it is clear that the living body is the proper subject of health and disease...since it pertains to natural philosophy to consider the living body and its principles, it must also consider the principles of health and disease.

...the study of health and disease is common to philosopher and physician. But since art is not the chief cause of health, but aids nature and assists it, it is necessary that the physician take from natural philosophy the more important principles of his science, as the navigator borrows from the astronomer. This is why physicians practicing medicine well begin from natural science.[39]

Since the time of Aquinas, of course, "natural philosophy" has developed into the whole panoply of physical and social sciences, as well as philosophies of nature and science. Yet the insight about general principles arising from the living body and returning through it to common features of human existence is important if we are to avoid the trap of making medicine responsible for all social causes of health and disease. An interdisciplinary and international effort can effect eradication of certain diseases. But medicine alone cannot assure human rights in health care. It must work with other disciplines, playing its restricted role in the clinical arena while articulating it efforts with politics, law, sociology, etc., to guarantee international human rights in health care.[40]

Clinical and Social Duties of Physicians and Other Health Professionals:
Conflict and Resolution

By *clinical medicine*, then, we mean the use of medical knowledge and skill for the healing of sick persons, here and now, in the individual physician-patient encounter. Clinical medicine so defined is the activity that defines clinicians qua clinicians and sets them apart from other persons who may have medical knowledge but do not use it specifically in clinical encounters, like the basic scientist or public health physician. Clinical medicine is the clinician's *locus ethicus*, whose end is a right and good healing action and decision for individual patients. Similarly, nursing at the bedside, dentistry, clinical psychiatry, social work, etc, each has its own *locus ethicus*.[41]

Moreover, clinical medicine is the instrument through which many public policies come to affect the lives of sick persons. No matter how broadly or socially-oriented we take medicine to be, illness will remain a universal human experience. Its impact upon individual human persons is the reason why medicine and physicians exist in the first place.[42]

Using clinical medicine as the paradigm for a philosophy of social medicine does not neglect the other branches of medicine, each of which has its own distinctive end. Thus, for basic scientists the end is the acquisition of fundamental biological knowledge of health and illness. This knowledge becomes a part of clinical medicine specifically when it is applied to the needs of a particular human being here and now. Similarly, preventive medicine has as its defining end, i.e., the cultivation of health and avoidance of illness. Hence, social medicine has its end in the health of the community or the whole body politic. When the knowledge and skills of any of the other branches of medicine are used in the healing of a particular person, then the ends of that branch fuse with the ends of clinical medicine. In clinical medicine, clinical nursing, etc., the good of the patient is the end, *primus inter pares*.[43] In social medicine, it is the good of society.

VII. SOCIAL RESPONSIBILITIES OF THE CLINICIAN

Throughout we have emphasized the primacy of the clinician's ethical responsibilities to be located in the good of his or her patient. Under what conditions may this responsibility be balanced by ethical obligations to the good of society?

First of all, in situations of natural disaster, just war, epidemics, and overwhelming emergency the clinician's knowledge and skill must be directed to the common good, the larger issue of social and community survival. Similarly, when a patient is a threat to the community, e.g. when the patient has a contagious disease and continues to place others at risk, the autonomy of the patient is no longer inviolable. Autonomy is limited when it results in the identifiable, probably, grave harm to others. The same is true of patients with HIV infection who refuse to tell their sexual partners, or an airline pilot, locomotive engineer, or crane operator whose condition poses a threat to the safety of others. In short, whenever the good of the patient, as perceived by the patient, poses a definable, grave, or probable risk to identifiable third persons, the physicians covenant with her patient is superceded by her duty to avoid a greater threat to third parties or to society at large.

In ordinary circumstances the physician's implicit promise to serve his own patient is a primary obligation. But within that obligation the physician is bound to consider societal good when both good ends can be sensed simultaneously. Thus, physicians are obliged to use the less expensive treatment if it is equally effective to the more expensive, even if there is some slight marginal benefit to the latter. Even more crucial is the obligation to avoid misuses, abuse, or overuse of treatments or diagnostic procedures. This is a violation of the obligation of competence, which requires the use of modalities of medicine that are effective, beneficial, and not disproportionally burdensome. In the long run, the best contribution the physician can make to conserve societal resources is to practice rational, effective, scientifically evaluated medicine. This happens also to be in the interests of the individual patient as well.

This does not mean that the physician should accept or assume the role of rationer or self appointed guardian of society's resources. To do so is to be in a morally unacceptable role of divided loyalty. Rationing should be explicit, not implicit. It should be determined by societal and institutional authority. The physician must still inform his patient about what is appropriate treatment. She must try by all legitimate means to obtain what is needed. But the final allocation of resources at both the micro and macro levels is a social not a professional decision.

When not joined in a covenant of trust with a particular patient, the physician has several obligations related to the common good. For example, physicians, nurses, and other health professionals are obliged to provide accurate, up to date unbiased technical information to policy makers, institutions, and administrators. They are required to avoid the kind of conflict of interest inherent in misleading exaggerations of benefit to advance one's favorite treatment procedure. Conversely, policymakers must be wary of one expert's depreciation of a competitors claims. The expertise of health professionals must be available as a sound factual basis for the decisions of policy and law makers. Without it political and economic considerations may distort good standards of care.

The requisite objectivity is difficult to achieve in our health care system that is, today, commercialized and market oriented to an alarming degree. Academic scientists and physicians are no less susceptible to self interest than their commercial counterparts. It is a rare research scientist who is totally free of conflict between his duties as physician and scientist, and his personal pursuit of self-interest, prestige, and power.

Yet, without reliable, verifiable, and accurate technical information, health care policy in the interest of the common good is impossible to design. Recovery of moral, as well as scientific, credibility has become a major task for today's health professionals. The moral high road is, of course, extremely difficult to follow. Without it, however the profession of medicine and the other health professions will lose whatever moral credibility they still retain.

Society, in the end, will be the loser. First, because it will be denied reliable technical information upon which to base public policy. Second, it will lose the example of one of the few remaining groups among which there is a substantial number who are dedicated to something other than their own self-interest. A society without an island, or two, of morally motivated professionals is a morally deprived society.

This takes us to the third level at which the professional may fulfill his or her societal duties and that is a member of a professional association or society. A medical or nursing association is defacto a moral community. Its members are united by a common public oath or commitment to act primarily for the benefit of those they purport to serve. They share in addition some set of moral precepts expressed in a moral code. Unless these moral dimensions are explicitly rejected, society assumes that they are the ethical signatures of the professions.

That professional societies today do not behave as moral communities does not erase the fact that their major ethical justification for existence is to advance the ends and purposes of the professions. If those ends and purposes are no longer moral in nature, professional associations become unions, guilds, or even the conspiracies against the public that George Bernard Shaw took them to be.[44] While not conspiracies, professional organizations today have become corporations, public relations agencies, and profit making organizations. Their size, capital holdings, and budgets are sometimes far in excess of what is required to function as moral communities, that is, as associations of professionals acting collectively to advance the purposes of medicine or the other health professions. Those purposes are focused upon the needs of sick persons or societies and not the propagation of self-interest.[45]

When they behave as moral communities, professional associations provide effective means whereby health professionals could fulfill their societal responsibilities. These associations should above all be advocates for the sick. They should act collectively via public education and political action to promote a just health care system, one in which the obligation of a good society toward its members to assure access, availability, and just distribution of health care could be realized.

Associations of health professionals have enormous latent moral power if only they choose to use it. They can influence public opinion, and raise public moral sensitivity to injustices, but only if they are genuinely acting for the good of society and not their own profit. We appreciate how far this notion of professional associations as moral communities is at present from the realities. However, as with all things in the moral realm, we can hope that what ought to be, may in fact, come to be.

Finally, health professionals can fulfill their social obligations as citizens. Here they can be advocates for what they believe to the elements of a just society or health care system. Here they can and, of course, will differ. It is here that they can express their own preferences apart from those of their fellow professionals. But here too, their votes and political participation should be guided by a sense of social good that transcends their own selfish self interest.

SUMMARY

Medicine has always existed within a social context in which the uses of medical knowledge and clinical decisions have impacted the good of society as well as the individual patient. In recent centuries the factual foundations for these interrelationships have been demonstrated. As a result, it has become clear that the social repercussions of medicine have serious ethical implications for both physicians and society.

We have, therefore, examined the relationship between the good of the individual patients and the common good in an effort to define a morally sound relationship between them, especially when they come into conflict. The proposition has been advanced that a philosophy and ethic of social medicine (or the medicine of society) is required that is comparable to the existing philosophy and ethic of clinical medicine. By comparing and contrasting the ethics and functions of clinical and social medicine some order of priority can be established when they come into conflict. The implications for clinicians in the partition of their ethical obligations to both patients and society are spelled out in terms of both an ethic of clinical and social medicine. While the physician is used as the example, the implications for all other clinicians are essentially the same, within the specific ethical framework of each profession.

NOTES

[1] Note: This paper was being written when Dr. Thomasma died unexpectedly. I have retained him as co-author although I have revised the text substantially. Nonetheless, I believe he would have no objection to the changes. His imprint on most of the text remains untouched.

[2] E.D. Pellegrino and D.C. Thomasma, *A Philosophical Basis of Medical Practice, Toward a Philosophy and Ethic of the Healing Professions* (Oxford: Oxford Univ. Press, 1981.

[3] E.D. Pellegrino and D.C. Thomasma, *For the Patient's Good: The Restoration of Beneficence in Health Care* (Oxford: Oxford University Press, 1987).

[4] E.D. Pellegrino, "The Internal Morality of Clinical Medicine: A Paradigm for the Ethics of the Helping and Healing Professions," *Journal of Medicine and Philosophy*, Vol. 26, No. 6, 2001, pp. 559-579.

[5] See Bernardo Ramazzini, *Diseases of Workers* (1713), rev., trans. Wilmer Cave Wright (Chicago: University of Chicago Press, 1940); Thomas Percival, *Medical Ethics: A Code of Institutes and*

Precepts Adapted to the Professional Conduct of Physicians and Surgeons (1803), ed. S. Russell (Manchester); Johann Peter Frank, *Complete System of Medical Polity* (System einer vollstandigen medicinishcen Polizey) (Manheim, Schwann, 1777-8) referenced in Fielding H. Garrison, *History of Medicine* (Philadelphia: W.B. Saunders Company, 1929), 321.

[6] Roy Porter, "Diseases of Civilization," in *Companion Encyclopedia of the History of Medicine*, eds. W.F. Bynum and Roy Porter, vol. I, 596-599 (New York and London: Routledge, 1993).

[7] Winch and Taylor.

[8] Aristotle, *Politics* 1330, b6.

[9] Ibid: a17.

[10] Aquinas, *De Regime Pricipum* 1, 15.

[11] Ibid: 1,1.

[12] E.D. Pellegrino, "Rationing Health Care: Conflicts within the Concept of Justice," in *The Ethics of Managed Care: Professional Integrity and Patient Rights*, eds. William B. Bondeson and James W. Jones (Dordrecht: Kluwer Academic Publishers, 2002).

[13] Especially social contractarians, like Veatch (2000).

[14] Emanuel 1999; A. Buchanan, "Managed care: Rationing without justice, but not unjustly," *Journal of Health Politics, Policy, and Law*, Vol. 23, No. 4 (1998), 687-95.

[15] (NE1094a1 and Book I)

[16] H. Jonas, *The Imperative of Responsibility: In Search of an Ethics for Technological Age* (Chicago: University of Chicago Press, 1984), 52.

[17] I. Murdoch, *Metaphysics as a Guide to Morals* (New York: Allen Lane, Penguin Press, 1993).

[18] Pellegrino, E.D. and D.C. Thomasma, *For the Patient's Good: The Restoration of Beneficence in Health Care* (Oxford: Oxford University Press, 1987).

[19] Veatch, 2000.

[20] Another example at this point would be the effort by Chinese society to limit the number of children its citizens could have for the sake of reducing its ever-burgeoning population versus the desire of individual citizens to bear children, often male heirs.

[21] R. Macklin, "Consent, coercion, and conflicts of rights," *Perspectives of Biological Medicine*, Vol. 20, No. 3 (1977), 360-71; United States Advisory Committee on Human Radiation Experiments, *Final Report* (Washington, D.C.: Supt. Of Docs, U.S. G.P.O., 1995).

[22] L.P. Knowles, "The lingua franca of human rights and the rise of a global bioethic," *Cambridge Quarterly of Healthcare Ethics*, Vol. 10, No. 1 (2001), 253-63.

[23] Pellegrino, E.D. and D.C. Thomasma, *For the Patient's Good: The Restoration of Beneficence in Health Care* (Oxford: Oxford University Press, 1987).

[24] S. Hauerwas, *Naming the Silences: God, Medicine, and the Problem of Suffering* (Grand Rapids, MI: Wm. B. Eerdmans, 1990).

[25] D. Novak, "The human person as the image of God," In *Personhood and Health Care*, D.N. Weisstub, D.C. Thomasma, and C. Herve (eds.) (The Netherlands: Kluwer Academic Publishers, 2001); D.N.

Weisstub and D.C. Thomasma, "Human dignity, vulnerability and personhood" In *Personhood and Health Care*, D.N. Weisstub, D.C. Thomasma, C. Herve (eds.) (The Netherlands: Kluwer Academic Publishers, 2001).

[26] E.H. Loewy, *Suffering and the Beneficent Community: Beyond Libertarianism* (Albany: State University of New York Press, 1991); *Freedom and Community: The Ethics of Interdependence* (Albany, NY: State University of New York: 1993).

[27] J. Welie, *In the Face of Suffering: The Philosophical-Anthropological Foundations of Clinical Ethics* (Omaha, NE: Creighton University Press, 1998).

[28] U. Jensen and G. Mooney, *Changing Values in Medical and Health Care Decision Making* (Chichester/ Wiley/New York, NY: A.R. Liss, 1990).

[29] T.K. Kushner and C. MacKay, "Joseph J. Jacobs on alternative medicine and the National Institutes of Health," *Cambridge Quarterly of Healthcare Ethics*, Vol. 3 (1994), 442-8.

[30] An example might be the practice of female castration in some African cultures. This practice has been widely criticized by bioethicists and physicians.

[31] E.D. Pellegrino and D.C. Thomasma, "Dubious premises – evil conclusions: Moral reasoning at the Nuremberg Trials," *Cambridge Quarterly of Healthcare Ethics*, Vol. 9, No. 2 (2000), 261-74; also see: M. Gross, "Treading Carefully on the Moral High Ground: Response to 'Dubious Premises – Evil Conclusions: Moral Reasoning at the Nuremberg Trials.'" *Cambridge Quarterly of Healthcare Ethics*, Vol. 10, No. 1 (2001), 99-102.

[32] R.G. Wilkinson, *Unhealthy Societies: The Affliction of Inequality* (London/New York: Routledge, 1996).

[33] D.C. Thomasma and E.D. Pellegrino, "Challenges for a philosophy of medicine of the future: A response to fellow philosophers in the Netherlands," *Theoretical Medicine*, Vol. 8 (1987),187-204.

[34] E. Van Leeuwen and G.K. Kimsma, "Philosophy of medical practice: A discussion approach," In *The Influence of Edmund D. Pellegrino's Philosophy of Medicine*, D.C. Thomasma (ed.) (The Netherlands: Kluwer Academic Publishers, 1997), 99-112.

[35] M. Foucault, *The Birth of the Clinic; An Archeology of Medical Perception* (London: Tavistock Productions, 1976).

[36] S.J. Reiser, *Medicine and the Reign of Technology* (Cambridge: Cambridge University Press, 1978); *The Machine at the Bedside* (Cambridge: Cambridge University Press, 1984); Jonas 1984. See also: R. Vos and D.L. Willems, "Technology in medicine: Ontology, epistemology, ethics and social philosophy at the crossroads," *Theoretical Medicine and Bioethics*, Vol. 21, No. 1 (2000), 1-7.

[37] R. McKeon (ed.), *The Basic Works of Aristotle* (New York: Random House, 1941).

[38] St. Thomas Aquinas, "Preface to the Commentary on Sense and the Sensed Object," Ed and Trans by R. McInerny, *Thomas Aquinas: Selected Writings* (London/New York: Penguin Books, 1998), p. 453.

[39] Ibid: 15-16.

[40] J. D'Oronzio (ed.), *Cambridge Quarterly of Healthcare Ethics*, Vol. 10, No. 3 (2001); D.C. Thomasma, and G. Diaz Pintos, *Autonomy and International Human Rights* (The Netherlands: Kluwer Academic Publishers, forthcoming).

[41] E.D. Pellegrino, "The internal morality of clinical medicine: A paradigm for the ethics of the helping and healing professions," *Journal of Medicine and Philosophy*, Vol. 26, No. 6 (2001), 559-79.

[42] See Hippocrates, *On the Art*.

[43] Some of the difficulty of the Hastings Center group in arriving at a consensus arose because these distinctions were not made clearly enough. The group tended to expand the definition of medicine so broadly as to absorb or "medicalize" almost all aspects of life. Such an expansion defeats any attempt to define ends. It places ends in conflict with each other and weakens any attempt to establish a hierarchy of goods among the many ends "medicine" may serve. See: I. Nordin, "The limits of medical practice," *Theoretical Medicine and Bioethics*, Vol. 9 (1999),105-23.

[44] G.B. Shaw, *The Doctor's Dilemma* (New York: Brentano's, 1913).

[45] E.D. Pellegrino and A.S. Relman, "Professional medical associations: ethical and practical guidelines," *JAMA*, Vol. 282, No. 10 (1999) 984-6.

ROSEMARIE TONG

TAKING ON "BIG FAT":
THE RELATIVE RISKS AND BENEFITS
OF THE WAR AGAINST OBESITY

ABSTRACT: Most health care ethicists are ill at ease in the world of public health ethics where the whole community or population is the object of focus, and the main moral tug-of-war is between the competing values of individual freedom and social welfare. Nevertheless, they are increasingly being pushed into the public health realm to address everyday issues, including overeating. I will first briefly note some of the health-status and health-cost concerns raised by the increased incidence of overweight and obesity among the American population. Next, I will discuss some of the most plausible explanations for Americans' expanding girth, including feminist Susan Bordo's analysis of consumer-capitalism. Then, using a framework developed by epidemiologist Geoffrey Rose, I will discuss the relative risks and benefits of both a "high-risk" (clinical) and "population" (public health) approach to obesity and overweight. Finally, I will conclude that the benefits outweigh the risks of a public health campaign against obesity.

KEY WORDS: health care ethicists, obesity, high-risk approach to obesity, population

Health care ethicists are very comfortable in the world of clinical ethics where the individual patient reigns supreme, and the principles of autonomy, non-maleficence, beneficence, and justice are routinely balanced against each other. However, most of them are rather ill at ease in the world of public health ethics where the whole community or population[1] is the object of focus, and the main moral tug-of-war is between the competing values of individual freedom and social welfare. Nevertheless, like it or not, health care ethicists are increasingly being pushed into the public health realm to address everyday issues such as smoking, drinking, and most recently, eating.

By using the methodological tools of epidemiology[2] public health officials have been able first to identify and then help control both tobacco-related and alcohol-related health problems in the United States. They have had their greatest success fighting smoking, which is probably the single best predictor of poor health and a premature death.[3] Today smoking is no longer as American as apple pie. The Lucky Strike girl who used to tout the pleasures of smoking on the 1950s television program, the *Hit Parade*, is a distant memory. Although U.S. smokers are by no means an extinct species, their number is dwindling, particularly in those states that have

M. Boylan (ed.), Public Health Policy and Ethics, 39-58.
© *2004 Kluwer Academic Publishers. Printed in the Netherlands.*

aggressively fought the tobacco industry. For example, in the 1980s about 50% of Californians smoked; today, as a result of two decades of anti-smoking campaigning, only 25% do. Moreover, if the present "smoking-is-bad-for-you" message continues to be loudly broadcast, Californian public health authorities predict that by the year 2010 only 10% of Californians will be lighting up.[4]

Despite some unsettling reports that the rate of smoking is increasing among certain populations such as young girls,[5] smoking rates have substantially decreased not only in California but also throughout the remaining forty-nine states. Education is among the chief means used to reduce smoking rates. Health care professionals from the U.S. Surgeon General down to the local school nurse have uniformly proclaimed the unmitigated health hazards of smoking. In addition, for fear of addicting children to nicotine before they are old enough to appreciate the health risks of smoking, the federal government has banned cigarette ads on television, and in other ways restricted the tobacco industry's ability to market its wares to young people (Joe Camel no longer freely roams the streets).[6] Moreover, state governments have imposed taxes on tobacco products and joined private citizens in successfully suing tobacco giants for their failure to inform the public about the true dangers of smoking. Finally, private as well as public institutions, facilities, workplaces, schools, and enterprises now restrict or forbid smoking in public places. Smoking is, for all practical purposes, no longer a sexy or sophisticated habit associated with being a controlled, cool person; rather it has become a dirty and nasty addiction associated with being an out-of-control, inconsiderate, and self-destructive person. Comment Dan E. Beauchamp and Bonnie Steinbock: "...the cigarette's long fall from the celluloid lips of Humphrey Bogart and Lauren Bacall to the chilled lips of office workers huddled outside buildings in the winter is a remarkable social transformation ushered in by more than the fear of lung cancer."[7]

Significantly, public health authorities have not had as much success fighting alcohol as they have had fighting tobacco, and this for several reasons. In the first place, the public remains largely convinced that there is a distinction between being a "true alcoholic" (who has a very big problem) on the one hand and being a social drinker (who has no problem) on the other hand. In general, forgiving folks view alcoholics as victims of their environment and perhaps even their genes.[8] In contrast, unforgiving folks view alcoholics as persons who knowingly drive while intoxicated, injuring or even killing innocent others in automobile collisions; or as persons who use their alcoholism as an excuse to engage in domestic violence or provoke bar-room fights. Yet for all their disagreements about how to view "true alcoholics," both forgiving and unforgiving folks agree that social drinkers are responsible individuals who supposedly know "when they've had enough."

Another reason for alcohol's slow entrance into the world of public health is the alcohol industry's ability to distinguish between "hard liquor" on the one hand and "fun beverages" on the other. Because the public views lite beers and wine coolers as harmless,[9] it is hard to stigmatize all alcohol as "bad for you." A final reason why alcohol has not become "Public Health Enemy Number Two" (assuming that smoking is "Public Health Enemy Number One") is that the health care community itself is quite divided about the relative health risks and, more recently, alleged health benefits of some types of moderate drinking. Several studies have found that a glass or so of wine a day actually keeps the doctor away.[10]

Given the public health system's experience with alcohol-related as opposed to tobacco-related health problems, then, it seems somewhat quixotical for the U.S. Surgeon General's Office to target overweight and obesity as its next object of conquest. As much as people like to smoke and drink, they like to eat even more, so much so that U.S. public health authorities may find their Waterloo in the belly of the American citizenry. Moreover, despite the obvious advantages of using public-health as well as clinical strategies to combat overweight and obesity, public-health strategies may also have some disadvantages. In order to bring these risks into focus, I will first briefly note some of the health-status and health-cost concerns raised by the increased incidence of overweight and obesity among the American population. Next, I will discuss some of the most plausible explanations for Americans' expanding girth. Then, using a framework developed by epidemiologist Geoffrey Rose,[11] I will discuss the relative risks and benefits of both a "high-risk" (clinical) and "population" (public health) approach to obesity and overweight. Finally, I will conclude that despite the considerable risks of a public health campaign against obesity, its benefits are somewhat greater. Prudence argues in favor of adding measured public health strategies to clinical strategies in the struggle to improve American's health status.

I. OBESITY AND OVERWEIGHT: HEALTH-STATUS CONSEQUENCES AND HEALTH-COST IMPLICATIONS

On December 14, 2001 former U.S. Surgeon General David Satcher announced that the American "obesity epidemic" had become so extensive that eating too much of the "wrong" things would soon rival smoking for the position of the number one cause of preventable deaths in the U.S.[12] Using new National Heart, Lung, and Blood Institute (NHLBI) guidelines on obesity and overweight to measure the American population's general corpulence,[13] public health officials determined that 61 percent of U.S. adults aged 20 to 74 are either overweight (34 percent) or obese (27 percent),[14] and that 13 percent of children aged 6 to 11 and 14 percent of adolescents aged 12 to 19 are overweight or worse.[15] Supposedly any adult with a Body Mass Index (BMI) of 25 kg/m^2 through 29.9 kg kg/m^2 is overweight, and any adult with a BMI of 30 kg/m^2 or greater is obese.[16] In addition, any child or adolescent who has a BMI at or above the 95th percentile for his or her relevant sex- and age-specific group is overweight or worse.[17] Translated into pounds, the language most of the public speak, individuals over 20 percent of their ideal body weight (as set by government standards) are merely "overweight," while individuals over 30 percent of their ideal body weight are obese.[18]

Although government public health officials emphasize that the health risks for obese individuals are far greater than those for overweight individuals,[19] by no means do they give merely overweight individuals a clean bill of health. Rather they suggest that many of the health risks specifically associated with being obese are to a lesser extent also associated with being overweight. These risks include premature death, type 2 diabetes, heart disease, stroke, hypertension, gallbladder disease, osteoarthritis (degeneration of cartilage and bone in joints), sleep apnea, asthma, breathing problems, cancer (endometrial, colon, kidney, gallbladder, and postmenopausal breast cancer), high blood cholesterol, complication of pregnancy, menstrual irregularities, hirsutism (presence of excess body and facial hair), stress incontinence (urine leakage caused by weak pelvic-floor muscles), increased surgical risk, psychological disorders such as depression, and psychological difficulties due to social stigmatization.[20] Economists estimate that in the year 2000 alone obesity, and to a lesser extent overweight, cost the U.S. health care system $117 billion ($61 billion in direct costs and $56 billion in indirect costs).[21]

II. OBESITY AND OVERWEIGHT: EXPLANATIONS FOR THE EPIDEMIC

Most explanations for the "fat" epidemic are elaborated versions of the simple truism that Americans eat too much and exercise too little. The U.S. Surgeon General's Office simply comments "[f]or the vast majority of individuals, overweight and obesity result from excess calorie consumption and/or inadequate physical activity."[22] Food is everywhere, but so are Americans' recliners. Too many of us sit both at work and at play.

Among the more compelling explanations for Americans' growing girth is one provided by Greg Critser, a health care economist. As he sees it, Americans are fat because "being fat—at least so far—makes economic sense."[23] According to Critser, the American waistline was held in check prior to World War II by two economic forces: (1) the high cost of food, particularly restaurant and convenience food; and (2) the high expenditures of calories in both the paid and unpaid (domestic) workforce. People used to work up a sweat laboring, including laboring around the home; and they ate their meals at home at scheduled times. All day "gnoshing" or "grazing" was reserved for major holidays, if that. Subsequent to World War II, however, advances in technology and agricultural productivity threw these two economic forces into disequilibrium. Food, including so-called fast food became inexpensive and very easy to get, whereas exercise became very difficult, even costly to get. Today one has to find time to get to the Fitness Center which charges a hefty annual fee. Comments Critser:

> ...In the past, we used to get paid for expending calories; now, we get paid for being sedentary. It's part of the job description. "Demand" for weight is up. We, as the "sellers" of our weight, like it that way, and so we resist any attempt to alter our "market". On the supply side, ubiquitous "super-size" options provide an economy of scale for overconsumption.[24]

Critser's solution to this state of affairs is to make it easier to get exercise and harder to get food, a point to which we shall return later.

Although Critser's economic analysis helps explain why Americans are getting fat, cultural critic Susan Bordo provides a deeper analysis of why some Americans are seriously overweight (obese), while other Americans are seriously underweight (anorexic). According to Bordo, in advanced consumer-capitalism, the "contradictory structures of economic life" create an "unstable, agonistic construction of personality."[25] In other words, it is contradictory to expect Americans simultaneously to both produce more than they consume *and* to consume more than they produce. Viewed as producers, Americans' task is to work relentlessly, ignoring bodily cares for sleep and food; but viewed as consumers, Americans' task, is to "eat, drink, and be merry;" to multiply their needs, and then demand they be immediately satisfied. Comments Bordo, "...we find ourselves continually besieged by temptation, while socially condemned for overindulgence." [26] Whether we are producing or consuming, we find ourselves in the wrong, unable to get things right. On one extreme, anorexia becomes a metaphor for "an extreme development of the capacity for self-denial and repression of desire (the work ethic in absolute control);"[27] on the other extreme, obesity becomes a metaphor for "an extreme capacity to capitulate to desire (consumerism in control)."[28] Given these two extremes, bulimia, an endless, imprisoning cycle of binging and purging, becomes the metaphor for advanced consumer capitalism's "Golden Mean." Comments Bordo:

For bulimia precisely and explicitly expresses the extreme development of
the hunger for unrestrained consumption (exhibited in the bulimic's uncontrollable food hunger) existing in unstable tension alongside the requirement, that we sober up, "clean up our act," get back in firm control on
Monday morning (the necessity for purge—exhibited in the bulimic's vomiting, compulsive exercising, and laxative program.[29]

Bordo's analysis, even more than Critser's, suggests that in order to curb the obesity
epidemic and other eating disorders, as well as many of the other addictions and disorders that plague Americans, the structure of advanced consumer capitalism itself
must be combated. I very much doubt, however, that the U.S. Surgeon General's
Office is able and ready as well as willing to fight this particular Goliath.

III. OBESITY AND OVERWEIGHT: CLINICAL AND PUBLIC HEALTH STRATEGIES FOR THEIR PREVENTION AND/OR TREATMENT

Given some of the cultural explanations for the manifestation of behavioral extremes
among much of the U.S. population, and also the ways in which food fits perfectly
into a producer-consumer model, public health authorities are likely to have more difficulty controlling food than they had controlling tobacco and even alcohol. To the
degree that they are frustrated in their attempts to make overweight or obese
Americans healthier, however, public health authorities may be tempted to lobby for
restrictions that ultimately do at least as much harm as good. In order to see how the
war against Big Fat could go wrong, it is helpful to compare and contrast the two
strategies traditionally used to prevent a particular disease or cluster of health-related
problems: the "high-risk" approach on the one hand and the "population" approach
on the other.[30]

The High-Risk Approach to the Prevention and Treatment of Overweight and Obesity
Clinicians typically favor a high-risk approach to overweight and obesity. The clinician identifies a patient as "overweight" and counsels the patient that unless he gets
his weight under control, he is likely to become obese and suffer dire health consequences. To drive the point home, the clinician orders a variety of tests for the patient.
If the test results are positive for high cholesterol or high blood sugar, the clinician
shows the patient that he is already at risk for heart disease and/or diabetes and had
best trade in his morning plate of biscuits and gravy for a small bowl of unadorned
oatmeal. The chastised patient then leaves the clinician's office probably with a body
image problem as well as a health problem that requires some sort of prescription.

1. Advantages of the High-Risk Approach: According to epidemiologist Geoffrey Rose, there are at least five advantages to the high-risk approach to obesity, [31] a condition which is increasingly viewed as a disease per se and not simply as a predictor for other diseases. [32] First, the high-risk approach to obesity yields "individual-appropriate" interventions. For example, if a patient has a high-cholesterol reading, more than likely, the clinician will suggest to him some sort of low-fat diet (a point to which we will return) and some sort of aerobic exercise. If the patient returns six months later more corpulent than ever and with the same or even higher cholesterol reading, the clinician may elect to prescribe a drug to the patient—for example, one of the new cholesterol-reducing statins. In addition, the clinician may counsel the patient more aggressively about the need for proper diet and exercise, referring him to a diet and nutrition specialist for additional help. Should the patient come back yet another six months later, more fat than ever and begging for some sort of "diet pill," the clinician may prescribe an appetite suppressant. If still another six months pass, and the same patient comes back, this time morbidly obese and on the verge of a major coronary event, he may recount to the clinician a series of events that runs as follows: "Doctor, I went to your diet and nutrition specialist. She couldn't really help me. So I went to Weight Watchers. I lost weight, but as soon as I didn't have to go to those weekly "weigh-ins," I started to gain weight again. I tried the Ornish diet, the Atkins diet, the cabbage-soup diet, the liquid diet, etc., etc. None of them worked. Let's face it. I love to eat. But I do want to live. I don't want my "ticker" to stop. So maybe I need to go cold turkey and get that gastric-bypass surgery I keep hearing so much about..." And if the patient is now at least 100 pounds above his ideal weight, and has truly tried every other way to control his weight, the clinician may refer him to a bariatric surgeon who may elect to reduce the patient's stomach size by as much as 99 percent. [33] Within several months, the patient will no longer be obese for obvious reasons.

In addition to identifying interventions appropriate to the individual patient, the high-risk approach to preventing obesity has a second and third advantage; namely, subject motivation on the one hand and clinician motivation on the other. To the degree that the patient and clinician truly believe that the patient has a bona fide health problem that can be prevented (or if prevention fails, successfully treated), to that extent will both the patient and the clinician supposedly be motivated to do something about it.

A fourth advantage of the high-risk approach is its cost-effectiveness. Instead of wasting time and resources on normal weight individuals, the clinician focuses on already overweight individuals, particularly those who threaten to enter the obese category. Why waste money educating slim-enough people about the dangers of a condition they do not yet manifest? Instead, hammer the message about "fat" home to those patients whose clothes are already painfully tight.

The fifth advantage of a high-risk approach to obesity is that it has a favorable benefit-risk ratio. Because many studies show that staying out of the obese category and conceivably also the obese category and conceivably also the overweight category promotes good health, clinicians feel justified to recommend dietary restrictions, prescription drugs, or even gastric-bypass surgery to a patient. To be sure, the clinician knows some patients will take the diet-and-exercise message to the extreme and become underweight or even anorexic; and other patients will let the message go in one ear and out the other, trying the limits of clinicians' patience. Similarly, the clinician knows some patients will abuse their diet pills, taking too many of them; and others will find themselves victims of diet drugs that may have been pushed to market too rapidly. For example, the prescription diet drug Fen-Phen resulted in many of its users developing thick or leaky heart valves which triggered such symptoms of coronary distress as shortness of breath, fatigue, chest pains, and abnormal heart rhythms; [34] and Fen-Phen's cousin drug, Redux, resulted in at least 123 deaths.[35] Finally, the clinician knows that a few patients (to be precise, 3 out of every 200) will die during bariatric surgery or shortly thereafter, usually due to infection from leaking sutures; and about 8 percent of bariatric-surgery patients will develop complications such as nearly permanent vomiting and ulcers.[36] However, for the most part, the clinicians knows most patients will not diet themselves down to prison-camp weight levels, be given harmful drugs, or die during stomach-reduction surgery. On the contrary, most patients' best interests will be well served in the course of their treatment for obesity.

2. Disadvantages of High-Risk Approach: Despite its benefits, like all strategies to prevent a disease or a cluster of health-related problems, the high-risk, clinical strategy has disadvantages as well as advantages. First, there are the predictable difficulties and costs associated with screening patients. Because the experts remain divided about which standard/scale/chart/index to use for determining whether an individual is at risk (i.e., over his/her "ideal" weight), the most prudent course of action would seem to require screening *all* patients, including ones that may be of ideal weight, and counseling *all* of them about proper diet and nutrition. However, such a massive screening program would be very costly for the health care system and time consuming for clinicians.[37] Moreover, many patients would be tempted not to schedule their annual physical exam for fear of being "put-on-the-spot" about their "beer bellies" and "saddlebags."

A second problem with the high-risk strategy is that it is palliative and temporary. In other words, the high-risk strategy targets individuals who are particularly susceptible to the underlying causes of obesity without seeking to alter anything over and beyond the individual's *personal* behavior or lifestyle. Still, there will always be "susceptibles" to obesity—or why, pray tell, spend so many research dollars hunting for the so-called obesity gene along with the biological and/or environmental cues that trigger it?[38] Prevention and control efforts would, as Geoffrey Rose points out, "need to be sustained year after year and generation after generation;"[39] and given that so many Americans seem to have either a genetic and/or environmental susceptibility to obesity, the second disadvantage of the high-risk strategy is hard to underemphasize.

A third disadvantage of the high-risk strategy is that *if* the real problem is not overweight or obesity per se but the other health-related problems to which these conditions are linked—e.g., cardiovascular disease—, then the high-risk strategy is of limited potential for the individual and the population. In other words, most overweight individuals and many obese individuals with risk-factors for coronary heart disease will remain well for many years, even for their whole lifetimes; while a significant number of ideal weight people will experience some form or another of coronary heart disease. To add specificity to this general point, epidemiologist Rose comments:

> This point came home to me only recently. I have long congratulated myself on my low levels of coronary risk factors, and I joked to my friends that if I were to die suddenly, I should be very surprised...The painful truth is that for [a man in the lowest group of cardiovascular risk] the commonest cause of death—by far—is coronary heart disease! Everyone, in fact, is a high-risk individual for this uniquely mass disease.[40]

In sum, overweight is not a sufficient condition for developing certain diseases any more than ideal weight is a sufficient condition for not developing them. Moreover, the further irony *is* that, from a population perspective, "*a large number of people at a small risk may give rise to more cases of disease than the small number who are at a high risk.*"[41] Still, if the majority of the U.S. population (i.e., the 61% overweight/obese American) are indeed all at high-risk for certain dread diseases, then irony may not prevail. If the number of high-risk individuals is actually higher than the number of small-risk individuals, *more* high-risk individuals than low-risk individuals will at some time manifest dread disease *x* or dread disease *y*.

Significantly, the fourth and final disadvantage of the high-risk strategy to overweight and obesity is probably the one that serves to push these two related conditions into the center of the public-health arena. Specifically, the high-risk strategy is "behaviorally inappropriate." In other words, it does not change the factors that cause so many people to overeat. Comments Rose:

> ...Eating, smoking, exercise and all our other life-style characteristics are constrained by social norms. If we try to eat differently from our friend it will not only be inconvenient, but we risk being regarded as cranks or hypochondriacs. If a man's work environment encourages heavy drinking, then advice that he is sort damaging his liver is unlikely to have any effect. No one who has attempted any of health education in individuals needs to be told that it is difficult for such people to step out of line with their peers. This is what the "high-risk" strategy requires them to do.[42]

Glossing over the fact that in contemporary American culture, the odd person is probably the person who is not on a diet—at least some of the time—Rose's general point is correct. Unless certain social norms, economic forces, and political pressures change, most behavioral imbalances will remain in place.

The Population Approach to the Prevention and Treatment of Overweight and Obesity

1. Advantages of the Population Approach: According to Rose, the chief advantages of a population approach to a health-related problem are that it is "radical," has a "large potential" for the population, and is "behaviorally appropriate." Unlike the high-risk approach, which is content to alleviate or remove the signs and symptoms of a health problem, nothing other than eliminating the underlying cause of the problem—in this instance, the problem of overweight/obesity—will satisfy the population approach. Comments Rose:

> If nonsmoking eventually becomes "normal," then it will be much less necessary to keep on persuading individuals. Once a social norm of behavior has become accepted and (as in the case of diet) once the supply industries have adopted themselves to the new pattern, then the maintenance of that situation no longer requires effort from individuals. The health education phase aimed at changing individuals is, we hope, a temporary necessity, pending changes in the norms of what is socially acceptable.[43]

For this reason alone, it is easy to understand why public health authorities tend to focus their energy on children who are obese and/or overweight.[44] If we raise a generation of children who prefer a 60-minute exercise session to a 60-minute video game session, and who actually like to eat moderate portions of fruits, vegetables, and grains more than super-sized French fries and cheeseburgers, then we will supposedly have generation after generation of adults who eat primarily or only healthy foods.

2. Disadvantages of the Population Approach: So obvious are the advantages of a population approach to the obesity/overweight epidemic that it is easy to overlook the disadvantages of this approach. First, a population strategy is, as Rose observes, highly susceptible to the "Prevention Paradox," according to which "A preventive measure which brings much benefit to the population offers little to each participating individual."[45] In other words, unless a presently overweight/obese individual is personally convinced that he/she is at a *serious* risk for a feared disease, it is unlikely that that individual will make dietary or physical activity changes. If the diet, fitness, and fashion industries have not been able to motivate the bulky Americans to get thin, it is probably unrealistic to think that health educators will succeed where advertising magnets have failed. Americans cannot quite decide just how much (or how little) they really want to be thin and/or healthy. They continue to spend 129 billion dollars per annum to consume super-sized fast food,[46] while simultaneously spending 35-50 billion dollars per annum on diet foods, diet drugs, exercise aides, and the like.[47] If Americans truly long to be as thin as models or fit as athletes, so much so that they would consider aborting a fetus with the "obesity gene,"[48] why, then, are they, as a population, getting fatter?

Significantly, patients' lack of real motivation to change their diet and exercise patterns are often coupled with clinicians' lack of motivation to take a population-approach with their overweight/obese patients. Some clinicians are tired of dealing with weight recidivists—that is, the notorious "yo-yo" dieters who lose weight only to regain it and then some. Other clinicians are reluctant to badger patients about their "weight problem" either because they themselves are "well-rounded" or because they do not think that being overweight is that much of a health risk, even if being morbidly obese is. Still other clinicians are loathe to get into the diet-and-exercise counseling business, largely because healthcare insurers do not generally reimburse physicians for time spent "merely talking" to patients.

A final disadvantage of a population approach to a health-related problem is the benefit-risk ratio. Rose reminds us that "[i]n mass prevention each individual has usually only a small expectation of benefit, and this small benefit can easily be outweighed by a small risk."[49] In all fairness, however, this truism may not apply in the case of a population approach to obesity/overweight. Without any apparent risk to themselves, it would seem that all individuals could benefit by maintaining an ideal weight. Nevertheless, I am unconvinced that forcefully aiming the entire American population towards the elusive goal of "Ideal Weight" is entirely risk free. Fighting food and the food industry in the same way, say, that tobacco and the tobacco industry were fought may not be as harmless to the general public as a prima facie inspection of the population approach suggests.

IV. REASSESSING A PUBLIC HEALTH (POPULATION) APPROACH TO OBESITY/ OVERWEIGHT: REASONS FOR CAUTION

A. Hidden Medical Rights of Waging "War" Against Obesity/Overweight

Before identifying the *non-medical* risks of a population approach to obesity/overweight, it is important to consider the *medical* risks of such an approach. The first medical risk of a population approach to overweight/obesity is rooted in the ongoing debate within the medical community about what constitutes ideal weight. When the National Heart, Lung, and Blood Institutes (NHLBI) issued its 1999 guidelines on obesity and overweight, millions of Americans "went from being 'okay' to 'overweight' overnight."[50] Although many health care professionals unquestionably accepted the new weight standards, others challenged them as unnecessarily labeling far too many individuals as overweight (e.g., Brad Pitt and Michael Jordan)[51] or obese (e.g., Sammy Sosa and Arnold Schwarzenegger),[52] thereby giving them reason to view themselves as unhealthy and/or unattractive. As critics see it, U.S. weight experts may have set the standard for ideal weight too low because they, like most other people in the U.S., are affected by the cult of thinness, according to which one can never be too thin. To the degree that U.S. weight experts' *medical* views are value laden, however, they may be inappropriately linking thinness not only to attractiveness but also to health. The value-ladenness of one's views about "ideal weight" manifests itself in the fact that the 1918 "body beautiful," Annette Kellerman, weighed 137 pounds at 5 foot 3 fl inches (overweight by today's standards), whereas one of today's "body beautifuls," Kate Moss, weighs 97 pounds at 5 foot 6 fl inches (underweight, one would hope, even by today's standards).[53] Is the true ideal of health better approximated by a woman more like Moss or more like Kellerman? In an effort to answer this question, it is useful to note that when the NHLBI issued its 1999 guidelines, it chose to ignore the second National Health and Nutrition Examination Survey (NHANES II), which used higher BMI cutpoints, and labeled only half as many individuals as overweight.[54] Moreover, recent studies indicate that the individuals added to the overweight category by the NHLBI standards are at no more risk for mortality than ideal weight individuals are, and that being severely underweight is probably less healthy than being slightly overweight.[55]

Also overlooked, this time by the public, is that the NHLBI's primary goal was supposedly not to get overweight people down to their ideal weight, but to ensure that overweight people did not become obese.[56] So much for caveats and good intentions, however. As soon as the newly-labeled overweight masses heard they had been declared fat as a base fiddle (or that's how it sounded to them), they were primed to beg their clinicians for aid-in-dieting. By then the American Cancer Society[57] and the Nurses Health Study[58] had published their respective studies, both of which concluded that even slightly overweight individuals were primed for a premature visit from the Grim Reaper, and neither of which mentioned health problems associated with being underweight, including metabolic abnormalities and osteoporosis.

Not only is it debatable just how much a mortality and morbidity risk being overweight (as opposed to obese) actually is, it is also debatable what constitutes a healthy diet—i.e., a diet that prevents obesity and overweight. Although it seems as if the U.S. Department of Agriculture's (USDA) food pyramid dates back to the Depression or World War II, it actually dates back only a decade or so.[59] The design of the food pyramid is supposed to visually convey the federal government's recommendation for a healthy diet. The tiny tip of the pyramid represents fat and sugar. The next tier down consists of high protein foods, including dairy products, meat, fish, nuts, and dried beans. The next to last tier on the widening pyramid is chock full of vegetables and fruits, and at the broad base of the pyramid is carbohydrate heaven: breads, cereals, rice, and pasta.

Food pyramid not withstanding, Americans have grown much fatter since 1992, the year of the food pyramid's initial publication. Critics of the food pyramid suggest that, as defined by the USDA, "healthy eating" may be partly to blame. The nation's food supply is about 50 percent larger than the population's need for food. Because the various segments of the food industry, each of which has a place on the food pyramid, need Americans not only to eat but to eat a lot, they spend $30 billion a year trying to get Americans to overconsume[60] and, for the most part, the federal government is complicit in their effort. Specifically, the USDA assigns each food group on the pyramid a number of recommended servings per day, ranging from "use sparingly" for fat and sugar to 6-11 servings for bread, cereal, rice, and pasta.[61] However, the USDA chooses not to highlight the size of these portions (e.g., a serving of pasta is fi cup, not the typical 3 cup serving of pasta most restaurants serve). Portion size is located in the fine print of the food pyramid's penumbra. Perhaps the federal government thinks citizens should know that a *portion* of bread equals a slice, not a load of bread. Then again, say the critics, the federal government, like the food industry, may be focused on the bottom line. Unless Americans keep on consuming, they won't need to do much producing in the future as the nation's economy grinds to a near halt.

Taking swipes at the food pyramid and the various vested interests it may or may not protect is a growing sport among various camps of "diet doctors" and their followers. On the one hand, disciples of Dr. Dean Ornish eat lots of carbohydrates, no added fats (such as oils and nuts), and no red meat.[62] On the other hand, disciples of Dr. John Atkins indulge in steak, eggs, and butter (all high fat)—but avoid potatoes, toast, and cereal (all high carbohydrate). [63] Neither group has any kind words for each other or their common foe: sugar. [64] Then there are the vegetarians. They join together in their opposition to meat, but subdivide themselves into vegans (who avoid all animal products, including leather shoes, for example), ovo-vegetarians (who eat vegetables and eggs), and pesco-vegetarians (who eat fish but no other animal). [65] Be they adherents of Ornish, Atkins, or some form of vegetarianism, each group maintains their diet is the healthiest one or, in the case of some vegetarians, the "holiest" one.

V. CONCLUSION

Over and beyond the fact that both the notion of ideal weight and a healthy diet are contestable is the fact that, truth be told, most Americans favor a quick-fix approach to overweight/obesity over a slow-and-steady approach. They prefer the Atkins diet to the Ornish diet because it helps them to lose pounds more swiftly, even if somewhat unsafely.[66] They prefer dangerous appetite suppressants or metabolism boosters over self-imposed small portions of food and exercise regimens. They even prefer, as we noted above, bariatric surgery if it solves their weight problem once and for all. Moreover, most Americans want to have it both ways: to eat super-sized meals that end with hot fudge sundaes *and* have lean bodies. They want weight-loss miracles such as the ones typically described in popular magazines. For example, in the weekly women's magazine *Women's World*, readers were informed about a research study on oleylethanolamide (OEA), a stomach chemical that suppresses the appetite and from which pharmaceutical companies may be able within the next five years to create a "drug that 'tricks' your body out of its cravings."[67] Readers were also informed that the stop-smoking medication bupropion (Wellbutrin and Zyban) has the unexpected, common side-effect of weight loss;[68] and that researchers are working hard to develop AOD964, a synthetic chemical, which will supposedly boost patients' metabolism and increase their fat-burning capabilities.[69] The ultimate dream is to find the "obesity gene" mentioned above, and either curb its "bad behavior" or eliminate it entirely from the gene pool.

Americans' love of the "quick-fix" is not the only problem, however. Most Americans care more about being thin than being healthy; and included in "most

Americans" are a significant number of health care authorities who regard their fat patients as weak and out of control. Health care authorities often overemphasize the problem of overweight and underemphasize the problems of underweight, which include poor nutrition, psychological harm, and obsessions with food and eating. [70] To be sure, there are currently more overweight than underweight Americans. However, precisely because our culture is one of extremes, caught in the consumer-producer yo-yo of advanced capitalist forces, an all-out war on corpulence could conceivably result in millions of Americans becoming too thin instead of too fat. But, as anyone who has given the subject of being too thin knows, an anorexic, food-phobic population is no more healthy than an obese, food-obsessed population.

To be sure, it is not public health authorities' intention to make Americans food phobic, and so far the "war" on obesity and overweight has been waged in a restrained manner. Indeed, the war has resulted in nothing more radical than the kind of legislation Senators Bill Frist and Jeff Bingman proposed in August 2002. Frist and Bingman's "Improved Nutrition and Physical Activity Act" or IMPACT would provide approximately $258 million in grants to train health care practitioners in treatment of obesity and overweight, fund nutrition education in the schools, subsidize walking and bike paths, and provide tax incentives to employers that structure employers' workplace and workday to allow for fitness breaks. [71] Other common-sense legislation includes bills to ban soft drink machines and "empty calorie" machines in elementary schools (and also perhaps high schools) as well as to provide school meals that conform to the thinking that built the USDA food pyramid. If the reason that not only the soft-drink industry but also teachers have balked at such bills is that the soft drink companies give schools a percentage of the proceeds from the sale of their products, and that schools are in desperate need of these monies to provide students with computers, physical education, and other "frills," then the solution is for the government to make sure that schools have enough funds not to need Pepsi's or Coke's money. It is contradictory for the government to fret about increasingly rotund children without being willing to pay for precisely those school programs that are most likely to help children develop a healthy lifestyle. Banning "goodies" from school premises is of little value when it is unaccompanied with meaningful health education programs which, by the way, should stress the dangers of being underweight as well as being overweight, [72] and which point out that most high-fashion models, jockeys, gymnasts, etc. are not necessarily healthy despite that fact that they look glamorous and/or can get their bodies to do amazing things. Too thin is not any more healthy than too fat.

Also appealing are proposals to include nutrition education, weight management, and obesity treatment in health care insurance packages, as are proposals to include treatment for eating orders which are often more costly to treat than obesity. Less appealing are proposals to require eating establishments to prominently list all the calories and nutrients contained in the foods they serve. The absolute extreme of this would be to have Surgeon General's warnings on foods high in sugar, fat, or, if one is a follower of Dr. Atkins, in carbohydrates. Making people view food as toxic in the same way that they increasingly view tobacco as toxic does not strike me as resulting in psychological health. It would be a shame if eating an ice cream cone in public came to bring stares of disapproval, pushing people to eat their goodies in the privacy of their closets.

The worst possible way to fight "Big Fat" is, however, through lawsuits. Fortunately, lawsuits filed by obese Americans against fast-food companies such as McDonald's, Burger King, Wendy's, and KFC for making them fat have so far failed. For example, in 2003 a Manhattan federal judge dismissed a lawsuit brought against McDonald's by two obese teenagers who claimed that McDonald's did not provide sufficient information about the health risks linked to its burger and fries meals. The judge voiced the opinion that everyone knows that "junk food" is fattening, and that no one, including the plaintiffs is forced to eat at McDonald's "except, perhaps parents of small children who desire McDonald's food, toy promotions or playgrounds and demand their parents' accompaniment."[73] Perhaps the greatest problem with lawsuits against fast-food chains, other than their frivolous nature, is that they will not lead Americans to develop life-long strategies for eating well. If anything, they will lead gullible people to believe that so long as they do not eat fast-food, they will live long in good health.

Let's face it. People are eating big portions of "unhealthy foods" in their homes as well as at restaurants, and our weight problems are unlikely to go away unless we get a grip on the culture that drives us to excess in every aspect of our lives. Simply put, a "bulimic" culture that requires people to alternate between binging (consuming) and purging (producing) runs on fat. Healthy eating requires a healthy culture and, regrettably, our culture, for all its achievements and successes is not a healthy culture.

NOTES

[1] Beauchamp, Dan E. and Bonnie Steinbock, eds. *New Ethics for the Public's Health*. (New York: Oxford University Press, 1999), p. 25.

[2] Mosher, James F. and David H. Jernigan. "New Directions in Alcohol Policy." *New Ethics for the Public's Health*. Eds. Dan E. Beauchamp and Bonnie Steinbock. (New York: Oxford University Press, 1999), p.138.

[3] Tobacco Information Prevention Source. *Centers for Disease Control and Prevention*. (18 July 2002) < http://www.cdc.gov/tobacco/mmwr4903fs.htm>.

[4] Shute, Nancy. "Kicking the Habit." *U.S. News & World Report*. (17 December 2001): 48.

[5] Wroe, David. "Smoking Out the Reasons Why Teenagers Take It Up." *The Age* (Melbourne). (3 June 2002): Features:11.

[6] Glantz, Leonard H. "Controlling Tobacco Advertising: The FDA Regulations and the First Amendment." *New Ethics for the Public's Health*. Eds. Dan E. Beauchamp and Bonnie Steinbock. (New York: Oxford University Press, 1999), p.164-176.

[7] Beauchamp and Steinbock, 133.

[8] Mosher, James F. and David H. Jernigan. "New Directions in Alcohol Policy." *New Ethics for the Public's Health*. Eds. Dan E. Beauchamp and Bonnie Steinbock. (New York: Oxford University Press, 1999), p. 137.

[9] *Ibid.*, 144.

[10] "Moderate Drinkers Healthier Than Abstainers and Ex-Drinkers." *Health & Medicine Week*. (19 November 2001 - 26 November 2001): 3.

[11] Rose, Geoffrey. "Sick Individuals and Sick Populations." *New Ethics for the Public's Health*. Eds. Dan E. Beauchamp and Bonnie Steinbock. (New York: Oxford University Press, 1999): 28-38.

[12] Neergaard, Laureen. "American Weighs Obesity Epidemic." *Toronto Star*. (14 December 2001): 28.

[13] National Institutes of Health (NIH), National Heart, Lung, and Blood Institute (NHLBI). Clinical Guidelines on the Identification, Evaluation, and Treatment of Overweight and Obesity in Adults. (Washington, DC: HHS, Public Health Service, 1998), p. xxiii.

[14] National Center for Health Statistics, Centers for Disease Control and Prevention. *Prevalence of Overweight and Obesity Among Adults: United States*. (31October 2001). <http://www.cdc.gov/nchs/products/pubs/pubd/hestats/obese/obse99.htm>.

[15] *Ibid.*

[16] National Institutes of Health, National Heart, Lung, and Blood Institute: 1.

[17] National Center for Health Statistics (NCHS), Centers for Disease Control and Prevention. *CDC Growth Charts: United States*. (31 October 2001). <http://www.cdc.gov/growthcharts/>.

[18] "Experts Address Obesity and Related Health Consequences." *Obesity, Fitness & Wellness Week* (25 August 2001 - 1 September 2001): 17-18.

[19] Calle, Eugenia E., Michael J. Thun, Carmen Rodriguez, et al. "Body Mass Index and Mortality in a Prospective Cohort of U.S. Adults." *New England Journal of Medicine*. 341 (15) (October 7, 1999): 1097-1105.

[20] National Institute of Diabetes and Digestive and Kidney Diseases. *Statistics Related to Overweight and Obesity.* (1 August 2002). <http://www.niddk.nih.gov/health/nutrit/pubs/statobes.htm>.

[21] Hellmich, Nanci. "Weighing the Cost: Is Being Fat an Illness That Should Be Covered By Insurance?" *USA TODAY.* (21 January 2002): 1D.

[22] Surgeon General David Satcher, *The Surgeon General's Call to Action to Decrease and Prevent Overweight and Obesity.* (Washington, DC: U.S. Department of Health and Human Services. 13 December 2001).

[23] Critser, Greg. "Modern Advances Make Obesity Too Easy." *Charlotte Observer.* (5 May 2002): 8D.

[24] *Ibid.*

[25] Bordo, Susan. "Reading the Slender Body." *Unbearable Weight: Feminism, Western Culture and the Body.* Ed. Susan Bordo. (Berkeley: University of California Press, 1993), p. 199.

[26] *Ibid.*

[27] *Ibid.*

[28] *Ibid,* p. 201.

[29] *Ibid.*

[30] Rose, 32.

[31] *Ibid.*

[32] "Weight Loss: IRS Recognizes Obesity as Disease That Qualifies For Tax Deduction." *Obesity, Fitness & Wellness Week.* (4 May 2002): 18.

[33] Contreras, Joseph and David Noonan. "The Diet of Last Resort." *Newsweek.* (10 June 2002): 46-47.

[34] Franklin, Robert. "Mayo Study Says Heart Problems Ease Over Time After Fen-Phen." *Star Tribune.* (2 December 1999): 2A.

[35] Willman, David. "How a New Policy Led to Seven Deadly Drugs." *The Los Angeles Times.* (20 December 2000): 1A.

[36] Contreras and Noonan, 47.

[37] Blumenthal, David, Nancyanne Causino, Yuchiao Chang, *et al* "The Duration of Ambulatory Visits to Physicians." *Journal of Family Practice.* 48, no. 4 (April 1999): 264-271.

[38] "Scientists Discover New Obesity Gene." *USA Today* (19 September 2000), website update.

[39] Rose, 34.

[40] Rose, 35.

[41] *Ibid.*

[42] *Ibid.,* 35-36.

[43] *Ibid.*

[44] See, for example: Shannon Brownlee, "Too Heavy, Too Young. *Time.* (21 January 2002); Wright, Charlotte M. and Louise Parker, *et al,* "Implications of Childhood Obesity for Adult Health: Findings from Thousand Families Families Cohort Study." *British Medical Journal.* 323, issue 7324 (1 December 2001): 1280-1284; and Elliot, Victoria Stagg. "Adult Options for Childhood Obesity?" *American Medical News.* 27 (May 2002): 27-28.

[45] Rose, 36.

[46] Horovitz, Bruce. "Fast-Food World Says Dive-thru is the Way to Go." *USA Today.* (3 April 2002): 1A.

[47] McNamara, Mary. "Ideas, Trends, Style and Buzz." *Los Angeles Times.* 22 (November 2000): 2.

[48] Strong, Carson, *Ethics in Reproductive and Perinatal Medicine* (New Haven, CT: Yale University Press, 1997), p. 138.

[49] Rose, 37.

[50] Lyons, Pat. "Great Weight Debate." *American Fitness.* 17, no. 1 (January/February 1999): 40-44.

[51] Parker, Kathleen. "He'll Gladly Sue You Tuesday for a Burger Today." *The Charlotte Observer.* (1 August 2002): 15A.

[52] *Ibid.*

[53] Garner, David M. et al. "Cultural Expectations of Thinness in Women." *Psychological Report.* 47 (1980): 483-491.

[54] Itallie, TB Van. "Health Implications of Overweight and Obesity in the United States." *Annals of Internal Medicine.* 103 (1985): 983-988.

[55] Strawbridge, William J., Margaret I. Wallhagen, and Sarah J. Shema. "New NHLBI Clinical Guidelines for Obesity and Overweight: Will They Promote Health?" *American Journal of Public Health.* vol. 90, issue 3 (March 2000): 340.

[56] *Ibid.*

[57] Calle, Thun, Rodriguez, *et al*, 1098.

[58] Manson, JoAnn E., Walter C. Willett, Meir J. Stampfer, *et al* "Body Weight and Mortality Among Women." *Journal of the American Medical Association.* 257 (1995): 353-358.

[59] Ness, Carol. "Our Shapes May Alter Pyramid." *Charlotte Observer.* (17 July 2002): E1.

[60] Winter, Greg. "America Rubs Its Stomach, and Says Bring It On." *The New York Times.* (7 July 2002): 5.

[61] Ness, E3.

[62] Kolata, Gina. "Scientist at Work: Dean Ornish; A Promoter of Programs to Foster Heart Health." *The New York Times.* (29 December 1998, late edition – final): F6.

[63] Taubes, Gary. "What if Fat Doesn't Make You Fat?" *The New York Times Magazine.* (7 July 2002): section 6: 20.

[64] *Ibid.*, and Kolata, 6.

[65] Corliss, Richard. "Should We All Be Vegetarians?" *Time.* (15 July 2002): 48-56.

[66] Taubes, 20.

[67] Korn, Wendy. "Weight-Loss Miracles in *Your* Future!" *Woman's World: The Woman's Weekly.* (16 July 2002): 14-15.

[68] *Ibid*, 15.

[69] *Ibid*, 15.

[70] Strawbridge, Wallhagen, and Shema, 341.

[71] Spake, Amanda. "A Fat Nation." *U.S. News & World Report.* (19 August 2002): 47.

[72] Strawbridge, 342.

[73] Parker, 15A.

DERYCK BEYLEVELD *and* SHAUN D. PATTINSON

INDIVIDUAL RIGHTS, SOCIAL JUSTICE, AND THE ALLOCATION OF ADVANCES IN BIOTECHNOLOGY

ABSTRACT: In this article, the authors apply Alan Gewirth's Principle of Generic Consistency. They argue that a defensible approach to medical research must give proper place to the problem of allocating any consequent advances in accordance with the moral rights and status of all agents. When this is done, they contend that some advances that cannot be given to everyone ought not to be available to anyone and certain types of applied medical research should not be conducted. This suggests that, when seen in the context of scarcity of resources, situations where the benefits of medical research can legitimately override the rights of research subjects are far more limited than is often recognised.

KEY WORDS: medical research, resource allocation, rights, Gewirth, privacy, confidentiality

Advances in biotechnology can create new treatment options, but only at a cost, and it is simply not possible for all potentially beneficial medical services to be made freely available to all. Additional demands on limited resources can only be met by devising ways of increasing available resources or diverting money from elsewhere (within or outside of healthcare), both of which have limitations. As a result, resource allocation difficulties are an inescapable part of healthcare provision and the possibilities presented by modern biotechnology are likely to increase these difficulties.

Competing moral theories offer different approaches to the difficulties posed by the need to allocate medical services in the context of limited public resources. This paper will briefly explore four prominent moral positions. The moral theory of the American philosopher Alan Gewirth will be presented as the most tenable.[1] It will be argued that medical research cannot be regarded as an unqualified good. To be morally justified, medical research must be scientifically efficacious, respect the moral worth and interests of all research subjects (including their rights to privacy and confidentiality), and be directed towards goals that are compatible with the moral worth and interests of all. It will be argued that some treatment services cannot be delivered in a way compatible with the moral worth and interests of all. More specifically, there are some treatment services that, when they cannot be offered to everyone, ought not to be available to anyone. Thus, contrary to the prevailing regulatory orthodoxy, this

M. Boylan (ed.), Public Health Policy and Ethics, 59-72.

will require the prohibition of certain types of applied research and the removal of the concomitant incentives.

MORAL FRAMEWORKS AND RESOURCE ALLOCATION

The sheer variety of potential criteria for distinguishing the morally permissible from the morally impermissible makes presenting an overview of competing moral frameworks a difficult task. There are, however, a number of prominent camps or positions on the issue of just allocation of advances in biotechnology. We will briefly explore four types of moral theories: utilitarianism, virtue ethics, libertarian rights ethics, and welfare rights ethics.

Utilitarianism is a collection of moral theories rejecting rights-based theories in favour of achieving the best possible balance of utility over disutility. The most popular version is preference utilitarianism,[2] which seeks to maximise the subjective preferences of persons in a calculus where all preferences count equally. For preference utilitarianism, rights-language can be no more than a convenient shorthand indicating that a certain state of affairs will contingently maximise preferences. Thus, whether or not there is a "right" to a medical treatment or service is contingent upon the consequences of recognising such a right on persons' preferences.

Virtue ethics also rejects rights-based theories; indeed, such theories reject all action-based moralities in favour of character-based values. For virtue ethics, of which feminist "care ethics" is a version,[3] all moral obligations are linked to human flourishing (assessed according to some "objective" criterion). Thus, whether or not there is a "right" to a medical treatment or service will depend on whether such a right is necessary for the flourishing of the individual for whom the service is to be provided. According to such a view, moral judgments should focus on character and motivational traits, rather than actions and results.

There are two variants of rights-based moral theories: libertarian rights ethics and welfare rights ethics. The former recognises only negative rights (rights to non-interference with the object of the right) and rejects all positive rights (rights to assistance with regard to the object of the right). Accordingly, a strict libertarian rights theorist[4] would reject rights to have any medical treatment provided by the state. Indeed, levying the taxation required for state medical provision is a *prima facie* violation of the (negative) rights of individuals. In contrast, a welfare rights ethics would recognise both positive and negative rights.[5]

For completeness, it should be noted that there are two conceptions of rights in rights-based theories. The will (or choice) conception holds that the benefit of a right is waivable by the rights-holder, which is rejected by the benefit (or interest) conception.

This dispute has consequences for what or who can be a rights-holder.[6] Since both conceptions distinguish negative and positive rights, the distinction between libertarian and welfare ethics applies whether or not the benefit conception is linked to a rights-based or duty-based ethics.

The moral theory to which we adhere, that of the American philosopher Alan Gewirth,[7] is a welfare rights ethic where rights are understood in terms of the will conception. Its supreme principle—the "Principle of Generic Consistency" (PGC)—grants both positive and negative rights to all agents, where agents are understood as beings with the necessary capacities to be intelligible subjects and objects of practical precepting. Agents are, thus, beings who are able to voluntarily pursue and reflect upon their chosen purposes. The rights granted to agents are rights to the "generic conditions of agency", which are the necessary means of acting at all and acting successfully, irrespective of the specific purposes being pursued. The generic conditions and corresponding rights are ranked according to their "needfulness for action".[8] Those conditions required for acting at all ("basic" goods), take priority over those needed for acting successfully—within which category those required for maintaining one's level of purpose-fulfilment ("nonsubtractive" goods) take precedence over those required for advancing one's level of purpose-fulfilment ("additive" goods).

The PGC's opposition to a libertarian rights ethics derives, essentially, from its recognition of positive rights to the generic conditions. Positive rights are, however, limited by two other provisos.[9] The first of these is the *own unaided effort proviso*, which states that agents' positive obligations are limited to situations where the potential recipient is unable to secure the generic conditions by its own unaided efforts. The second proviso is the *comparable cost proviso*, which states that agents only have duties to aid other agents to secure their generic features when doing so does not deprive the aid provider of the same or more important generic capacities, as measured by the degree of needfulness for action.

These features of Gewirthian moral theory are consequences of the PGC's justification.

WHY PREFER GEWIRTHIAN THEORY?

The Dialectically Necessary Argument
According to Gewirth, agents must accept and act in accordance with the PGC on pain of contradicting that they are agents. In consequence, the PGC is, as Gewirth

expresses it, "dialectically necessary" for all agents: any and every agent must accept it because it follows with necessity (purely logically) from a premise that no agent can coherently deny, namely that it is an agent. Gewirth's reasoning may be presented as follows.[10]

If an agent has a goal (E) that it is motivated to pursue, and it is necessary for the agent to have something (X) in order to achieve E, then E gives the agent a reason to pursue X. If there are generic conditions of agency, then the agent categorically (i.e., simply by virtue of being an agent) has an *instrumental* reason to pursue/defend having the generic conditions *whatever E might be*. In short, for an agent to have the generic conditions is categorically in the agent's interests *as an agent*, and it follows that agents, *within the context of their agency,* must recognise that they categorically ought to pursue/defend having the generic conditions or contradict that they are agents.

It is equally clear that, because it is categorically in the agent's interests as an agent to have the generic conditions, it is categorically in its interests as an agent that others not interfere with its having the generic conditions and, indeed, that others assist it to secure or defend having them when it is unable to do so by itself. Thus, agents must (on pain of contracting that they are agents) consider that others categorically ought not to interfere with their having the generic conditions *against their will* (and, indeed, that others who can do so categorically ought to assist them to secure the generic conditions when this assistance is needed *and wished for*).[11] These other-referring ought judgments must be upheld by the agent simply because this non-interference or assistance is required to satisfy the criterion of the agents' interests as an agent to which *it* is necessarily committed.

To uphold these other-referring ought judgments is, however, to say that the agent has both negative and positive rights to the generic conditions under the will-conception of rights. Thus, all agents must (as agents) consider that they *themselves* have both positive and negative rights (under the will-conception) to the generic conditions.

But why must agents consider that *other agents* also have these generic rights? The answer is definitely not that *other agents* have as good a justification for asserting *their* rights as *I* (any particular agent) have for asserting *my* rights, and this is because their justification is *their* agency interests (their need for the generic conditions of agency), whereas my justification is *my* agency interests (my need for the generic conditions of agency), and the argument has not yet shown (as it must eventually) that agents must (on pain of contradicting that they are agents) guide their own actions by the interests of others in addition to their own. However, contrary to what some critics have alleged,[12] it nevertheless follows purely logically that agents who contradict that they are agents by not considering that they have the generic rights also contradict that they are agents if they do not consider that all other agents have the generic rights. This is because if I contradict that I am an agent by not considering that I have the generic rights then I must (in order to avoid contradicting that I am an agent) accept that I cannot uphold any position that implies that I can be an agent but not have the generic rights. However, it follows from this that I must not merely hold that I have the generic rights; I must also hold that I have the generic rights for the self-sufficient reason that I am an agent.[13] But it follows purely logically from "The fact that I am an agent is by itself sufficient for me to have the generic rights" that "The fact that X is an agent is by itself sufficient for X to have the generic rights". Consequently, agents contradict that they are agents if they do not consider that all agents have the generic rights, which is to say that the PGC is dialectically necessary.

Deriving Acceptance of the PGC from the Acceptance of Morality
Gewirth relies entirely on the dialectically necessary argument for the PGC. The reason why he does so is that, like Kant before him, he correctly appreciates that a moral principle can only be established *as* categorically binding by showing that it is connected entirely *a priori* with the concept of being an agent (with the consequence that conformity with the principle is revealed as a necessary condition of all rational action).

It is possible, however, to construct dialectically contingent arguments for the PGC, which are contingent in that (unless the dialectically necessary argument is sound) their premises can be denied by agents without self-contradiction.

One such argument may be directed at agents who accept the idea that there are morally binding requirements on action, defined as categorically binding *impartial* ones (i.e., categorically binding requirements that require the agents to take equal account of the interests of all agents in determining what they themselves may do).[14]

By virtue of accepting that there are categorically binding requirements on action, I (any agent) must accept that there are requirements that are binding on my actions simply because I am an agent, which entails that I must accept that there are ends that I ought to pursue on pain of contradicting that I am an agent (or as Kant would have said, I must accept that pure reason is practical). However, because there are generic needs, it follows that I must accept, regardless of what ends I might think are categorically binding on me, that I ought to possess the generic needs as means to my pursuit of these ends. Similarly, other agents must consider that they ought to possess the generic needs as means to their pursuit of any ends that they might consider to be categorically binding. In this sense, I must accept that the possession of the generic needs is in the interests of myself and all other agents. From this it follows, given my commitment to the idea that I must take impartial account of the interests of all agents, that I must accept that all agents ought to have the generic needs as means to whatever they consider to be categorically binding ends. But this is equivalent to my having to accept that all agents have claim rights to the generic needs under the will-conception of rights (which means that I must accept the PGC).

This argument is very much facilitated by, and is relative to, my acceptance of the requirement of impartiality. However, it can also be demonstrated that anyone who accepts that there are categorically binding requirements on action must accept the PGC on pain of contradicting this acceptance, even when this acceptance does not include acceptance of the requirement of impartiality.[15]

As we have just seen, the acceptance that there are categorically binding requirements on action requires me to consider that I ought to pursue my having the generic needs as means to whatever ends I consider to be categorically binding on me. However, because the idea that an agent is categorically bound to do X is the idea that it is simply by virtue of being an agent that the agent is bound to do X, I must, on pain of *contradicting the claim that there are categorically binding requirements on my action,* hold that I contradict that I am an agent if I do not accept that I ought to pursue my having the generic needs as means to whatever I claim to be ends categorically binding on me, whatever these might be. However, as we saw in the dialectically necessary argument for the PGC, it follows *purely logically* from the proposition that I deny that I am an agent if I do not consider that I have the generic rights that I deny that I am an agent if I do not consider that the self-sufficient reason why I have the generic rights is that I am an agent. Since this inference is purely logical, it must follow *purely logically* from the premise that I deny that I am an agent if I deny that I have Q (where Q is any property), that I deny that I am an agent if I deny that the self-sufficient reason why I have Q is that I am an agent. Consequently, I must, on pain of contradicting my acceptance of categorically binding requirements on my actions, consider that I deny that I am an agent if I do not consider that my being an agent is the self-sufficient reason why I categorically ought to possess the generic needs as means to my pursuit of whatever ends I claim to be categorically binding on me. But, from this, it follows purely logically that I must (on pain of contradicting my acceptance of categorically binding requirements on my actions) consider that all other agents categorically ought to possess the generic needs as means to their pursuit of whatever they consider to be categorically binding on their actions (hence that all other agents have the generic rights under the will-conception of rights).[16]

CRITERIA FOR RESOURCE ALLOCATION

There are a number of potential criteria (variously conceived as competing or complementary) for allocating access to advances in biotechnology. These include:[17]
(a) ability to pay;
(b) individual need;
(c) cost per potential recipient;
(d) age of the potential recipient; and
(e) likely effectiveness of the treatment on the potential recipient.

For a libertarian rights ethics, all things being equal, ability to pay is the only legitimate criterion for allocating access, as other potential criteria imply the existence of positive rights. Thus, all things being equal, allocation under a libertarian ethic would be determined by the market subject to such minimal regulation as is necessary to protect the negative rights of individuals.

For preference utilitarianism, the appropriate criterion or criteria is to be determined by a preference maximisation calculus. A crude application of such a calculus suggests that, all things being equal, access to medical resources ought to be available to the largest number possible and favour those individuals likely to gain the greatest benefit for the longest time. Thus, all things being equal, access ought to favour the young and those recipients for whom the treatment would be most effective (i.e., criteria (d) and (e)). The application of this calculus to advances in biotechnology is particularly difficult because the cost of access to any such advance is likely to be particularly high (so that more preferences might be maximised by using public resources for other purposes) and, for some advances, allocating on the basis of ability to pay is likely to generate resentment on the part of the majority of the population who cannot afford to pay (so that preferences are not maximised by (a)). Thus, preferences are likely to be maximised by ensuring that where access to a specific service is likely to be very costly, not available to everyone, and likely to create large-scale resentment from those who fail to gain access, access ought not to be available to anyone.[18] Moreover, as will be detailed below, some scientific aspirations are likely to pit those able to take advantage of consequent opportunities against those who are not so able. All things being equal, preference utilitarianism cannot accept the fulfilment of the preferences of a minority at the expense of the preferences of the majority.

For virtue ethics, the appropriate criteria must ensure the flourishing of the individual, which will depend on the criteria for human flourishing. Some advances in biotechnology are likely to undermine the flourishing of individuals (e.g., it is plausible that acceptance of their finality entails that individuals ought not to seek indefinite extension of life, see below). Where a particular treatment does not itself undermine human flourishing, however, those most likely to benefit (assessed according to the development of virtues) would seem to have priority; and others should be encouraged to display virtuous acceptance of being denied these benefits. Thus, a crude application of virtue ethics would appear to point towards criterion (b).

These applications of a libertarian rights ethic, preference utilitarianism, and virtue ethics are unavoidably crude, as these are groups of moral theories rather than specific theories. However, since, in our view, the most compelling moral epistemology is presented by Gewirth's argument to the PGC, hereafter we will focus on the application of the PGC.

For Gewirthian theory, any criteria for allocation of medical resources must recognise the equal status of all agents as rights-bearers. Priority is determined by the level of generic need of potential recipients, not their number. In this sense the PGC is distinctly distributive, rather than aggregative. Since ability to pay is not distributed solely according to need and merit, (a) is unacceptable as a criterion for allocating resources to which all agents have equal claim. Similarly, since age (when not connected with the likely effectiveness of treatment) has no bearing on the potential recipient's generic needs, it is not a legitimate criterion for resource allocation. Accordingly, all things being equal, individual need (weighed according to the criterion of degrees of needfulness for action) is the most appropriate criterion for allocating scarce public resources. All things being equal, it follows that, as a first principle of allocation, priority should be given to those who face an immediate threat to their lives.

One of us has argued elsewhere[19] that there is a paradigmatic case of a medical treatment or service that, when it cannot be given to everyone, ought not to be given to anyone. This is indefinite life-extending treatment understood as treatment designed to extend the lifespan of the subject well beyond that of the current maximum for the human race so that it as close to an infinite existence as possible.

Life is the most basic generic feature (and hence the most important right) and, as such, all agents' lives are of equal worth. Consequently, all agents have an equal claim to treatment granting super longevity in the form of an indefinitely extended lifespan. However, the sheer expense of such a service would make it impossible to offer it to everyone. Yet equal status can only be ensured by universal access, universal denial, or completely random allocation. Consequently, the PGC cannot condone an aspiration to maximise bodily existence where such a service granting super longevity is not available for all. Such an aspiration involves discrimination between agents at the point where all human agents are most fundamentally equal. Moreover, under the PGC agents have an imperfect duty to internally inculcate (or a perfect duty to aspire to inculcate) such views and values as are necessary to ensure respect for the intrinsic worth and dignity of all other agents.

Thus, in principle, treatment aimed at granting longevity as close to indefinite physical existence as possible should not be available to anyone unless it can be given to everyone.

Before exploring the practical limitations of attempts to enforce such a principle, it is useful to look at other advances in biotechnology to which similar principles might apply. A striking feature of treatment services designed to extend lifespan indefinitely is that, if successful, these will create a substantial disparity between those who have access to such services and those who do not. Successful recipients will have an exponentially longer lifespan than all other humans. The result will be a class of super humans who will be better placed to dominate society than any other group. Given appropriate circumstances, such biologically elite individuals (and countries, companies, and other groups possessing such individuals) will have incomparable advantages in relation to accumulated wealth and knowledge. Thus, if biotechnology did render all body parts essentially renewable, so as to grant an indefinitely extended lifespan, those able to gain access to such a service would gain a massive competitive advantage over all other humans (unless, of course, lifespan extension did not involve a corresponding extension of mental youth—fears beautifully conveyed by the Struldbrugs in Jonathan Swift's *Gulliver's Travels).*[20]

The threat of artificially created biological elites is evoked by many techniques and technologies that are currently only science fiction hypotheses. Any treatment capable of giving one group of humans a substantial biological advantage at a cost beyond the means of any feasible universal distribution mechanism has the potential to create and perpetuate social inequalities. For example, attempts to create super intelligent humans will, if successful, inevitably create a sharp distinction between the haves and the have nots that can only be avoided by universal access or prohibition. Fears evoked by "designer babies" largely stem from the realisation that many biological differences are determined before birth and so prenatal manipulation has clear potential for creating divisions between humans. Perhaps, the ultimate consequence of unconstrained trait manipulation before birth will be a division of what is now human society into sub-species due to genetic modification of one group so that they become unable or unwilling to reproduce with the genetically disadvantaged.[21]

These predicted consequences are far from unrealistic. Throughout history humans have shown an irrepressible urge to identity themselves as part of select groups distinct from other groups. Xenophobia, racism, sexism, and other forms of prejudice placing undue value on morally irrelevant features and attributes, all involve a failure to recognise the inherent dignity of individuals in another group. The creation of additional dramatic biological divergences present more than the opportunity for additional prejudices, the very use of such technologies undermines the equal moral worth of all. Technologies creating biological elites could ultimately result in war, as the majority will feel (justly) deprived and the minority will seek to protect their advantage.

PRACTICALITIES AND MEDICAL RESEARCH

We have argued that, all things being equal, some potential advances in biotechnology must be available to all or available to no one. Even where *completely* random allocation of such advances in biotechnology is a practical possibility, it can perpetuate social divisions and create biological distinctions between recipients and other human agents.

We recognise, however, that enforcement of such a principle would require the prohibition of certain types of applied research and the removal of the concomitant incentives (e.g., removal of intellectual property protection for the results of such research). We also recognise the power of economic, social, and individual pressures in favour of allowing research that aspires to, *inter alia,* indefinitely extend human life or massively extend human intelligence. If biotechnology did, for example, render all body parts essentially renewable so as to grant an indefinitely extended lifespan (which is, admittedly, extremely unlikely), it would be well nigh impossible to prevent determined individuals gaining access to the technology. We live in a world in which researchers and those seeking access to advances in biotechnology can easily travel to less prohibitive jurisdictions. Unfortunately, global regulation requires global agreement. It is, therefore, doubtful that research capable of revealing the necessary knowledge could be stopped now, and it would be practically impossible to prevent all access to such a service if it were successful.[22] It follows that we must hope that science will fail to achieve its most ambitious aims. Nonetheless, such practicalities can, at most, convert what would otherwise be duties to achieve into duties to try.

Consequently, difficulties presented by potential future developments should, at least, lead to reconsideration of the popular idea that medical research is best left governed by minimal regulation. To be morally justified, all things considered, applied medical research must take full account of *all* predictable consequences *and* involve a justifiable process.

Appreciation of this is, however, obscured by a tendency to regard medical research as an unqualified good as well as by exaggerating the benefits that medical research is likely to yield. This combination facilitates mischaracterisation of the relative importance of the rights of individuals and the interests of society in general when these come into conflict. This is particularly problematic where the research subject was initially a medical patient.

The obligation to protect patients' privacy and confidentiality is, we suggest, a likely victim of an over enthusiastic focus on the benefits of medical research using patient information. Medical patients may, of course, exercise their free, informed will and release whatever personal information they wish for non-therapeutic research, subject only to the rights of others. The benefits of all PGC-recognised rights are waivable. It must be recognised, however, that patients are particularly vulnerable. The doctor-patient relationship invites trust and confidence that is necessary for the physical and psychological well-being of patients, but such power inevitably creates potential for abuse.

If the likelihood and magnitude of potential benefits of medical research are exaggerated, it is all too easy for patient rights to be undervalued. There are few legitimate reasons for utilising patient information for medical research without the patient's free consent. Indeed, the patient might have a very strong interest in ensuring that his or her information is not utilised for certain types of research or any research at all. Is it likely that a Catholic woman would willingly consent to the use of her medical data for the development of contraceptives, that a pacifist would consent to his medical data being used for the development of biological weapons, or that a Jehovah's witness would consent to his/her blood products being used for research purposes? Of course, the autonomy of the patient in relation to such matters is not an absolute value, and we would not suggest that there are not considerations that might override this value. In relation to this, it has been claimed by some cancer epidemiologists that any requirement to obtain consent from cancer patients for their data to be entered on cancer registries for research purposes is to value privacy (in its aspect as representing the value of autonomy to control the use of information about oneself) above the saving of lives.[23] However, while we would agree that privacy may be breached to save a life, this is only so if it is not possible to protect privacy and save the life. Since consent to the use of the data would make it possible to protect privacy and save lives (assuming that the research would lead to life-saving cures, which is often little more than speculation), it must be impracticable or impossible to obtain consent before the value of saving life can be claimed to conflict with the value of privacy. However, if the argument for impracticability is that it is too much of an inconvenience for doctors to obtain the consent then it is not privacy that should be seen to be in conflict with the value of life, but the convenience of doctors. On the other hand, if it is held that the impracticability is a consequence of the fact that some patients will not consent, then it must be recognised that, where it is desirable to have data on whole populations (as cancer registers generally intend) the relevant sample is the whole human race—in consequence of which a 100% sample can never be obtained. In our view, unless there is a *specific* reason for believing that a less than 100% sample will seriously bias any research results, impracticability can only be successfully made out in relation to data already held on patients where it is a genuinely serious matter to try to go back to patients (many of whom will already have died) to get consent. It cannot be made out for the prospective scenario where the patient is in front of the doctor. Nevertheless, in the UK, lobbying by cancer epidemiologists has led to regulations (The Health Service (Control of Patient Information) Regulations 2002) being passed under section 60 of the UK's Health and Social Care Act 2001 that permit data on cancer patients to be used for research without consent on the grounds that it is impracticable to get consent in both scenarios. This raises a suspicion that the British Parliament, in permitting these regulations to be passed, allowed an over optimistic view of the value of medical research to obscure clear thinking about the case at hand.

CONCLUSION

What might be called "science-worship" (i.e., the belief that all advance of knowledge, especially knowledge of human biology, is an unqualified good) is just as problematic as "science-hatred" (i.e., the belief that science is inherently evil).[24] A truly defensible balance between these extremes must be consistent with the moral rights and status of all agents. While, in general, this signifies only that great care must be taken to characterise conflicts between rights and interests of different parties accurately and that proportionality must be ensured in overriding one right or interest by another, we have argued that there are some cases at least where it means that some applied research should never be conducted and, indeed, ought to be prohibited.

NOTES

[1] See *Reason and Morality,* (Chicago: University of Chicago Press, 1978).

[2] See, e.g., R. M. Hare, *Moral Thinking,* (Oxford: Clarendon Press, 1981).

[3] See, e.g., Eva Feder Kittay, *Love's Labor: Essays on Women, Equality and Dependency,* (London: Routledge, 1998).

[4] See, e.g., Robert Nozick, *Anarchy, State and Utopia,* (Oxford: Basil Blackwell, 1974).

[5] See, e.g., John Rawls, *A Theory of Justice,* (Oxford: Oxford University Press, 1972).

[6] According to the will theory, only beings able to waive the benefits of rights (agents) can be rights-holders. For an example of the benefit conception, see Tom Regan, *The Case for Animal Rights*, (London: Routledge, 1988).

[7] See Alan Gewirth, *Reason and Morality.*

[8] See Alan Gewirth *The Community of Rights,* (Chicago: Chicago University Press, 1996), 45–46.

[9] For explanation and examples of the application of these provisos see Shaun Pattinson, *Influencing Traits Before Birth*, (Aldershot: Ashgate, 2002).

[10] No attempt is made here to provide a full analysis or defence of the argument. This has been done in Beyleveld 1991. For a shorter presentation that takes account of some other objections, see Beyleveld and Brownsword 2001, Chapter Four.

[11] It is because the agent must value the generic needs only instrumentally (albeit categorically so) that the agent cannot be prohibited categorically from waiving the benefits that exercise of the rights would confer. The rights are claimed for the possibility of action and successful action. There is no premise in the argument that agents must value their agency as such. Such a premise is an evaluation that would require independent justification that cannot be secured by the mere claim to be an agent.

[12] Most notably, Richard B. Brandt (1981, 39–40).

[13] See Gewirth's "Argument from the Sufficiency of Agency" (ASA) (1978, 109–110). See also, Beyleveld and Brownsword 2001, 75.

[14] See Beyleveld and Brownsword 2001, 91–92 for a fuller version of this argument.

[15] See Beyleveld and Brownsword 2001, 93–94.

[16] A further dialectically contingent argument for the PGC, which is relative to the acceptance of human rights is available. This argument has particular significance in the context of legal systems that accept the fundamental nature of human rights. For a presentation of this argument see, e.g., Beyleveld and Brownsword 2001, 79–86.

It must be emphasised that, strictly speaking, the PGC is a principle of agent rights, rather than of human rights as such. It does not, however, follow that those humans who appear to be non-agents (such as newborns and the severely mentally disabled) are the objects of moral worth and protection only indirectly, as a result of their relations with those humans who appear to be agents. We have argued elsewhere that, all things being equal, the PGC requires that all humans be treated as agents insofar as possible and consistent with the PGC's other requirements: see Deryck Beyleveld and Shaun Pattinson, "Precautionary Reasoning as a Link to Moral Action," in Michael Boylan ed., *Medical Ethics*, 39–53. (Upper Saddle River New Jersey: Prentice-Hall, 2000). We argue there that the need for a precautionary approach is driven by the categorical nature of the PGC and epistemic limitations on knowing whether or not a particular being is an agent under the conception of agency applicable to the PGC.

[17] This list of potential criteria is not intended to be exhaustive. We have not, e.g., listed the "quality adjusted life-year" (QALY), which seeks to allocate according to the cost of potential treatment services relative to the number of years of life expectancy gained by such treatment services where life years are discounted according to ill-health, disability, and other factors affecting quality of life. This process involves a calculation wherein an additional year of healthy life expectancy would count as one and an additional year of less than healthy life expectancy would count as less than one. Thus, priority would be given to those treatment services with the highest cost-per-QALY (i.e., cost per additional year of quality adjusted life expectancy). Although this criterion has utilitarian overtones, it has been supported from other moral positions, such as contractarianism: see, e.g., Peter Singer, John McKie, Helga Kuhse, and Jeff Richardson, "Double Jeopardy and the use of QALY's in Health Care Allocation." (1995) 21 *Journal of Medical Ethics* 144–150. However, since the QALY calculation involves a combination of many of the criteria we have listed–notably as (c), (d), and (e)—it is not addressed as an independent allocative criterion in this paper.

[18] In appropriate circumstances, preferences might also be maximised by random allocation, rather than no allocation, e.g., where allocative criteria giving everyone an equal chance of gaining access to treatment services is unlikely to generate resentment from those denied access *or* where the disutility of any resulting resentment would be outweighed by the utility gained as a result of satisfying the preferences of those who do receive treatment.

[19] See Deryck Beyleveld, "Individual Rights and Problems of Social Justice" (2003), in L. Honnefelder ed., *The Impact of Genetic Knowledge on Human Life,* Kluwer (in press).

[20] In Book Three, Gulliver is first filled with envy at the opportunities bestowed by the super longevity of the Struldbrugs, which quickly turns to pity when he discovers that the reality if one of extended old age, rather than extended youth.

[21] See, e.g., Lee M. Silver. *Remaking Eden: Cloning, Genetic Engineering and the Future of Humankind?*, (London: Phoenix, 1998), 1–13; 281–286. For comments on this suggestion see Pattinson, *Influencing Traits Before Birth*, especially 6.5.

[22] Support for our empirical claims can be found in regulatory responses to past and current biotechnological advances, particularly genetic and reproductive techniques. Global pressures in favour of embryo research, for example, are ensuring that even those countries prohibiting such research are becoming more favourably inclined towards it and its contribution to knowledge. See Shaun Pattinson, "Current Legislation in Europe." In Jennifer Gunning and Helen Szoke, eds., *The State of ART Regulation*, (Aldershot: Ashgate, 2002), 7–19.

[23] See Ann Dix "Cancer Experts Call for Action on GMC's Confidentiality Rules," *Health Service Journal* 2 November 2000. See also, Deryck Beyleveld "GMC is Protecting Patients' Fundamental Right to Privacy," *Health Service Journal* 16 November 2000 (letter).

[24] Others have also argued for moral constraints upon the quest for scientific knowledge. Boylan and Brown, e.g., argue that there are moral limits on what they term the "Principle of Plenitude" ("What can be known should be known"): see Michael Boylan and Kevin E. Brown, Genetic Engineering: Science and Ethics on the New Frontier, (Upper Saddle River, New Jersey: Prentice Hall, 2001), chapter 7. They specify these limits as applying to: "(a) instances in which the means of obtaining the scientific ends are immoral and (b) instances in which the ends themselves may clearly be seen to be involved in a larger action that is, itself, immoral": *ibid.,* 86.

ROSAMOND RHODES

JUSTICE IN ALLOCATIONS FOR TERRORISM, BIOLOGICAL WARFARE, AND PUBLIC HEALTH

ABSTRACT: Terrorism, biological or chemical warfare, and the public health response to threats and events are on everyone's mind. At the same time, any public policy response must be just. Yet, it is difficult to achieve justice in public health policy because there is neither a single consideration nor a simple formula for success. A variety of considerations can legitimately support good policy and a variety of cloaked and disguised hazards can intrude on the policy-making process and, in the name of justice, lead to indefensible results. For public health policies to be just, the reasons behind them must be the ones that reasonable people would find most compelling and most appropriate. Policies must reflect the choices that reasonable people would make and the priorities that reasonable people find most pressing. To achieve a just result, those who are responsible for creating public health policy need to be alert to the kinds of illegitimate considerations that can distort and pervert any policy. This paper discusses both the priorities for justice in public health policies designed to protect against and respond to terrorism and biological or chemical warfare and for the hazards that policy makers should try to avoid.

KEY WORDS: Justice, public health, terrorism, bioterrorism, chemical warfare, triage, research, vaccination.

On the beautifully clear and sun filled morning of September 11, 2001, just before voting in the New York City Primary elections, news of a plane crashing into the North Tower of the World Trade Center was announced on the radio. As I drove towards Manhattan after casting my ballot, newscasters reported a second plane crashing into the South Tower. It became instantly clear to everyone that these were acts of terrorism. We were under attack. All around me the behavior of New York City drivers immediately converted from the standard aggressive mode to a remarkably accommodating style, and in a dramatically uncharacteristic way cars yielded to make way for the emergency vehicles that were suddenly racing down the highway. Only one car surged to follow in their wake.

By the time I reached Mount Sinai, the hospital where I work on the upper east side of Manhattan, the side street on which the emergency entrance is located had been

M. Boylan (ed.), Public Health Policy and Ethics, 73-90.
© *2004 Kluwer Academic Publishers. Printed in the Netherlands.*

cordoned off, police were on duty, and health care workers stood on the sidewalk beside an empty stretcher. The institution was locked down, people were admitted to the complex by only one of the many routine entrances, and those who entered were screened through two ranks of security personnel. Members of the medical staff stood waiting in the huge I.M. Pei designed glassed atrium, surgeons in their operating greens, other doctors who had raced back to campus from elsewhere in their jogging and golf outfits, and clusters of medical students who congregated in small huddles.

As I later learned, rehearsed emergency measures were enacted that morning throughout the institution. A disaster plan had previously been developed and practiced so that everyone knew what had to be done. A significant number of beds in each intensive care unit (ICU) were emptied. All elective surgery was canceled. Every patient who could have been sent home, was discharged. Collection activities in the blood bank went into high gear, but they were only accepting O- donors.

As we all later learned, two additional planes were crashed by terrorists that morning, one into the Pentagon, the other in a Pennsylvania field. Medical institutions near each disaster site responded. Tragically, although we had expected and prepared for many patients, few came. Over the next three days, Mount Sinai had just eleven World Trade Center related admissions.

After the attacks, health professionals were also called upon for their advice in dealing with the aftermath of the disasters. The public needed to know about the health dangers posed by the material that had disintegrated to dust and was being carried around New York City by the wind. They needed to know about the toxicity of the fires that continued to burn at ground zero. They needed to know whether it was safe to go outdoors, how far away from the site to keep, who was most at risk, how to protect themselves from contaminants. Doctors were asked for answers and for advice on safety measures. But in this age of evidence-based medicine, there was no evidence of the toxicity of such tremendous quantities of pulverized concrete or incinerated computer parts and office furnishings. There were odors in the air, but there was no evidence about whether they were harmful and what problems they could cause. The best that our experts could say was that efforts should be made to keep the dust out of the air. So streets and vehicles leaving the site were hosed down, and those working at the site were outfitted with masks, buildings were vacuumed and mopped and air filters changed. Extrapolating from what was generally the case with impurities in the air, warnings went out that those with respiratory problems, children and the elderly should take special care, stay indoors, and keep out of the playgrounds.

Then in early October 2001, envelopes laced with anthrax (*bacillas anthracis*) began to arrive in the mail at offices in a number of U.S. cities. This was biological

terrorism. Cipro was dispensed, postal facilities, office buildings and homes were tested and decontaminated. Public health officials worried about attacks with other biological agents, such as small pox, and started to plan for dealing with the possibilities.

Since the autumn of 2001, public health contingency planning for terrorism and
biological warfare has continued. These efforts have proceeded on several levels, as
measures to protect society from the harms of future possible attacks, as plans to prepare society to respond to future terrorism, and as efforts to try to learn from our
experience in order to avert some harms in the future. Since then studies have begun
to collect data so as to assess the short- and long-term health effects of the exposure
on those who worked at the World Trade site and those who live or work nearby. This
data is supposed to be helpful in understanding the kinds of hazards that are caused
by various contaminants and in designing public health measures and medical interventions to avoid or combat dangers.

Although there has already been some debate about strategy (e.g., universal vaccination to protect against a possible smallpox epidemic versus ring vaccination in
response to an actual terrorist infectious disease attack) and about the allocation of
resources (e.g., which victims to benefit and how much, whether to allocate resources
for planning and to which plan, whether to allocate resources for research and what to
study) the principles that underlie these decisions have been universally assumed without contention. Implicit in this silent agreement are the presumptions 1) that everyone
knows the guiding principles of justice and 2) that those principles have the solid
endorsement of a broad majority of the population. In what follows, I will question
both presumptions. I will examine the principles that are most commonly invoked in
discussions of the just allocation of social resources. I will also explore their relevance
to public health policy for confronting terrorism and biological warfare. The exploration of these vivid and timely examples should shed some light on our general
approach to the just allocation of a society's limited public health resources.

MEDICAL MODELS FOR EMERGENCY AND
PUBLIC HEALTH RESOURCE DISTRIBUTION

Triage is the broadly endorsed approach for responding to medical emergencies. It is
the approach that was immediately adopted by health care workers on September 11,
2001 and its appropriateness has not been challenged in any of the literature that I
have encountered since then. Triage in domestic medical emergencies requires health
care professionals to make judgements about the likely survival of patients who need
medical treatment. Recognizing that some people have urgent needs (i.e., they will

die or suffer significant harm if not treated very soon) and that the resources available are scarce (i.e., supplies, facilities, trained personnel), patients are sorted into three groups and they are either treated, or turned away, or asked to wait according to their group classification. Those who are not likely to survive are deprived of treatment so that the available resources can be used to save the lives of those who are more likely to live. Those who are likely to die without treatment but who are likely to live if treated promptly are treated first. Those who are in need of treatment but who can wait longer without dying are treated after those who are urgently ill.

In medical emergencies, health care professionals deliberately disregard the concepts of giving everyone a fair equal opportunity to receive medical treatment and they also pointedly ignore relative differences in economic and social standing. Instead, they focus exclusively on the medical factors of urgency of need and the likelihood of survival. No one presumes to measure whether or not each patient has received a fair or equal share of available resources. No one stops to assess who is more or less advantaged and no one criticizes medicine for not attending to those differences.[1] In fact, the long tradition of medical ethics, dating back at least to the Hippocratic tradition, requires physicians to commit themselves to the provision of treatment based on need. Hence the ethics of medicine appears to require physicians' unequal treatment and also non-judgmental regard of patients' worthiness. These long-standing expectations have not changed. These commitments remain in tact irrespective of recent writing on the just allocation of medical resources, and they have not been eroded or transformed by the events of September 11th.

Biomedical research and public health policies typically focus on populations. Biomedical research attempts disconfirm hypotheses about predicted outcomes and thereby to develop facts about the response of organisms with certain common characteristics. With respect to human subject research, groups of people are selected for study because of some relevant biological similarities. Any knowledge gained from the process is useful to the extent that it is applicable to all of those who share the common condition.

Public health policies are also designed to have an impact on all, but only those, individuals who are similarly impacted by a disease- or a health-related condition. In deliberately focusing on one affected group or another, biomedical research and public health policies typically provide benefits to only the target group. The goals of biomedical research and public health are pointedly directed at everyone who might benefit from them. By looking back at outcomes, researchers attempt to develop generalizable knowledge about biological or psychological reactions. By looking toward the future, public health officials attempt to develop some generalizable approach to the prevention, reduction, or treatment of biological or psychological problems.[2] And as with medical triage in the emergency setting, biomedical research and public health have not been criticized for holding to these agendas.

FAIR EQUALITY OF OPPORTUNITY AND PRIORITARIANISM

Since 1971, many of the positions on justice espoused by philosopher John Rawls, first in *A Theory of Justice*[3] and later in *Political Liberalism*[4] and other works, have come to play a significant role in public deliberation about the criteria for justice in society. One Rawlsian concept that has received especially broad endorsement is his commitment to what he calls "fair equality of opportunity." The other principle that has been widely supported is the "difference principle," and people who have embraced some version of that principle now refer to such views as "prioritarianism." These principles exemplify Rawls's view of what a liberal political conception of justice should include.[5]

Rawls's two principles of justice provide "guidelines for how basic [political] institutions are to realize the values of liberty and equality" and assure all citizens "adequate all-purpose means to make effective use of their liberties and opportunities."[6] Together these principles specify certain basic rights, liberties, and opportunities and assign them priority against claims of those who advocate for the general good or the promotion of perfectionism (i.e., the best possible society).

Rawls himself does not extend his principles of justice to health and medical care. In fact, he specifically maintains that "variations in physical capacities and skills, including the effects of illness and accident on natural abilities"[7] are not unfair and they do not give rise to injustice so long as the principles of justice are satisfied. Yet, several prominent authors who write about justice and medicine discuss medical allocations according to Rawls's principles. They either extend Rawlsian concepts to medicine or use the allocation of medical resources to frame arguments against his positions.

According to Rawls's first principle, justice requires a liberal democratic political regime only to assure that its citizens' basic needs are met and that citizens have the means to make effective use of their liberties and opportunities. A just political regime must provide its citizens with "primary goods," that is the "all-purpose material means for citizens to advance their ends within a framework of the equal liberties and fair equality of opportunity."[8] Rawls's second principle regulates the basic institutions of a just state so as to assure citizens fair equality of opportunity. The first principle has priority over the second in that it requires political institutions to provide whatever citizens must have in order to understand and to exercise their rights and liberties. According to Rawls, his principles assure such basic political rights and liberties as: liberty of conscience, freedom of association, freedom of speech, voting, running for office, freedom of movement, and free choice of occupation. They also guarantee the political value of fair equality of opportunity in the face of inevitable social and economic inequalities.[9] Both principles, therefore, express a commitment to the equality of political liberties and opportunities.

In Rawls's account, the difference principle is the second condition of the second principle of justice. Recognizing that economic and social inequalities are an unavoidable feature of any ongoing social arrangement, his second principle expresses the limits on unequal distributions. He holds that equal access to opportunities is a necessary feature of a just society, and then, so as to compensate for eventual disparities and to promote persisting equality of opportunity, he calls for corrective distribution measures through application of the difference principle. As Rawls states the principle, "Social and economic inequalities . . . are to be to the greatest benefit of the least advantaged members of society."[10] In other words, governmental policies that distribute goods between citizens must be designed to rectify inequality by first advancing the interests of those who are otherwise less well off than their fellow citizens.

A CHALLENGE TO FAIR EQUALITY OF OPPORTUNITY

Norman Daniels has used the Rawlsian concept of fair equality of opportunity to argue that health care should be treated as a basic need.[11] He maintains that "[h]ealth care is of special moral importance because it helps to preserve our status as fully functioning citizens."[12] Daniels expects his claim to lead to the conclusion that a just society should provide its members with universal health care, including public health and preventive measures. Recognizing that a society's limited resources will limit the amount of health care that a society can provide, Daniels proposes normal species function as the benchmark for deciding which care to provide. He holds that health care that is likely to restore or maintain normal species function should be provided. Nothing has to be provided, however, for those who are already within the normal range. Furthermore, Daniels points to the many social determinants of health inequalities and invokes Rawls's difference principle to claim that a just society should provide the most health care to those who are most disadvantaged with respect to health.

Both Daniels and his critic, Julian Savelescu, would agree to count at least some medical services as "primary goods" so that they are "treated as claims to special needs." From Daniels's point of view, the allocation of health care resources should aim at equalizing social opportunity. Savulescu has argued, in opposition to Daniels, that health care policy should aim, instead, at providing everyone with a "decent minimum" of medical services.[13] Savulescu would give preeminence to policies that will save the most lives and thereby maximize everyone's chance of being saved so that each can pursue the kind of life that reflects personal values and goals.

Because ideas about justice and medicine are typically discussed singly in artificially isolated contexts, it is hard to notice when and how they clash. Yet, the post September 11th broad consensus on emergency triage, public health research, and public health policy provide an occasion to consider justice across a broad spectrum of medical contexts. This array of examples also serves as a challenge to Daniels's assumption that a fair process of deliberation would yield consensus on justice in health care based upon fair equality of opportunity.

For example, emergency triage provides neither equal shares of care nor equal opportunity for future social participation. On the contrary, the distribution of resources under triage supports Savulescu's view and gives everyone a better chance at living than could be had by an equal distribution of resources. Emergency triage allocates resources by taking everyone's prognosis and expected outcome into account, but individuals certainly get unequal lots.

Similarly, public health research sometimes has no impact on social participation. If it turns out that we never have another disaster similar to what occurred on September 11th, if we never again experience a catastrophe that creates enormous amounts of pulverized concrete and incinerated computers, research on their effects may never promote the social participation of anyone. Or, if the burdens of the interventions that the studies support turn out to be prohibitively costly (e.g., give up skyscrapers and computers), they will not be adopted and no one's fair equality of opportunity will be advanced. Public health research involves a quest for information that may or may not be useful. It also involves the direction of resources to the needs of the relatively few affected individuals. So, the standard of promoting fair equality of opportunity never quite fits. Many other uses of the resources that are more likely to promote fair equality of opportunity should have preference over public health research if that were the only consideration to be taken into account. Yet, the consensus in favor of such research suggests that other reasons support its broad endorsement.

Furthermore, while public health policies sometimes meet the standard of promoting fair equality of opportunity, sometimes they do not. In some cases (e.g., anthrax, small pox), interventions are advocated because they are likely to save more lives than some alternative plan. The huge amount of resources devoted to decontamination of post offices and office buildings after the mail-disseminated anthrax attacks, was widely accepted. But the clean-up policy had only a hypothetical and distant relation either to the avoidance of deaths or to the promotion of fair equality of opportunity.

As Ronald Green has noted, "Daniels's mistake . . . is trying to decide such matters by reference to a single consideration—and not necessarily the most important one."[14] Emergency triage, public health research, and public health policy seem to rely on principles other than, or in addition to, fair equality of opportunity.

A CHALLENGE TO PRIORITARIANISM

Prioritization stands in opposition to utilitarian approaches to the distribution of scarce resources. Utilitarian allocations aim at the maximization of an outcome over a population and they deliberately ignore the relational and relative differences between individuals. Utilitarians try to identify an objective standard for calculating consequences and then base policy decisions for everyone in the population on the same cost-benefit analysis. Prioritarian distributions aim at the identification of unwanted inequalities and the distribution of resources so as to compensate for or correct them. Prioritarian allocations reflect a concern for how individuals fair in relation to each other and they attempt to advantage those whose position is worse than others'.

Numerous papers in the bioethics literature address the conflict between prioritarian concerns and utilitarian cost-effectiveness analysis in the allocation of medical resources. For instance, Dan Brock,[15] Frances Kamm,[16] and David Wasserman[17] argue the merits of one approach over the other in a variety of vexing cases. They reflect on the difference between policies that will save the lives of some people or save an arm for some other people. They are concerned with whether public policies should provide a greater advantage to some who are already well off (e.g., save the lives of the able-bodied), or provide a smaller advantage to some who are worse off (e.g., save the use of an arm for a group with some other pre-existing disability). These "tragic choices" discussions sometimes focus on identifiable individuals, but sometimes not, they sometimes address trade-offs of future significant harms against present small harms or more certain imminent harms against more hypothetical distant harms. Typically, these discussions favor policies that will allocate resources to immediate needs over future needs and benefits to identifiable individuals over benefits to those who cannot be currently identified.

Emergency triage policies for medical care prioritize patients in a very different way than what would be required prioritarianism. Biomedical research and public health policy also appear to follow principles that conflict with prioritarianism. The stunning absence of objections to these practices raises serious questions about the value and scope of prioritarianism. 1) Does the absence of objections and the apparent general endorsement of the principles that guide emergency triage, biomedical research, and public health policy constitute counter-examples to Rawls's "difference principle"? 2) Or, does the broad consensus on these allocations of medical resources suggest that terrorism and biological warfare present legitimate exceptions to prioritiarianism? 3) Or, is prioritarianism, as a general conception of justice, the result of a mistaken extension of Rawls's view? In other words, 1) do the wide agreement on medical triage and public health responses to terrorism and biological warfare constitute a refutation of the difference principle and, therein, suggest that Rawls stepped outside of the "overlapping consensus" of reasonable and rational views when he committed himself to the "difference principle"? 2) Or, is the "difference principle" an appropriate conception of justice that has to accommodate a few well-justified exceptions? 3) Or, does the "difference principle" have a narrower scope and do less work than prioritarians would imagine?

1) The principles of utilitarian equality and efficacy support the well-accepted views on emergency triage. When the time constraints of an emergency and the needs for medical resources significantly outstrip the available resources, responses based on equality and efficacy will certainly oppose prioritarianism. If both sets of principles (utilitarian equality and efficacy on the one hand, prioritarianism on the other) apply to the same kinds of issues, they cannot both be appropriate for guiding these allocation decisions. The intuitions that support the view that priority should be given to the least advantaged are undermined by the strong sense that relative differences should not come into play in decisions about emergency responses to terrorism and biological warfare. Questioning the commitment to prioritarianism in medical triage invites questions about the appropriate framework for policy decisions about public health needs and setting the research agenda.

This conflict between principles, coupled with the strong intuition in favor of utilitarian equality and efficacy in emergency triage also raises the question of whether Rawls's commitment to the "difference principle" is a necessary feature of his construction. Is it a principle that would surely be endorsed by reasonable representatives behind the veil of ignorance, or is it merely an expression of Rawls's personal intuitions and his own comprehensive views of morality and politics?

Rawls offers the "difference principle" in the context of choosing principles for the construction of a political framework for society. He is concerned with the guarantee of fair value for political liberties, that is, allowing each citizen roughly equal access to public facilities and the political process. He recognizes that the accumulation of personal wealth and the pooling of resources in collaborations of the wealthy for the promotion of their self-serving ends could threaten the political participation of those with lesser means.[18] He also appreciates the importance of limiting disparities in private property because large differences could undermine self-respect and provoke envy. These are crucial concerns for Rawls in the construction of a fair procedural process for the structure of society's basic institutions because ignoring these concerns would tend to threaten social stability.[19] Rawls's recognition of the need for stability in *Political Liberalism* makes the avoidance of destabilizing conditions a paramount political consideration. Stability provides the justification for the "difference principle" and it explains why the "difference principle" is essential to Rawls's political conception of justice as fairness.[20]

2) Although Bernard Gert, Charles Culver, and K. Danner Clouser do not subscribe to Rawlsian prioritarianism, in their book, *Bioethics: A Return to Fundamentals,* they have taken the position that medical emergencies justify exceptions to the rules of ordinary morality.[21] In the context of this discussion, their approach can raise the question of whether terrorism and biological warfare are exceptions to prioritarianism because they are emergencies. According to the analysis of Gert and his colleagues, the importance of saving a life to most people, can make imminent death and the likelihood of preserving life adequate reasons for overriding the standard constraints on causing pain or limiting liberty. But, in their discussions of justifiable paternalism, Gert and his colleagues are concerned with justifying the infliction of some loss of freedom or harms of short duration on a patient for the sake of providing the same patient with what appear to be more significant and enduring benefits. In decisions about emergency triage and the allocation of funds to public health research and policy, however, the issue is not about how to benefit a particular individual (i.e., the inherent tension between commitments to beneficence and respect for autonomy), but about whether saving more lives adequately explains allowing exceptions to the "difference principle" (i.e., distributive justice). The concerns that justify exceptions for the emergency treatment of an individual are significantly different from the concerns that might justify deviations from prioritariansm in emergencies.

Furthermore, before something can be found to be a legitimate exception to a rule, the place of the rule must be established. Once a rule is accepted for governing a particular activity, violations of the rule have to be well-justified before they can be accepted as legitimate exceptions. When it comes to the allocation of medical resources, however, we have a real question about whether prioritarianism, or utilitarian equality and efficacy, or some other principles should rule. And in situations of emergency triage, biological warfare, and public health research or policy the reigning intuitions tend to oppose prioritarianism. These facts transform the question into the problem of putting the cart before the horse: before we can decide that something is a legitimate exception to the rule, the rule must be settled.

3) As Rawls explains it, the "difference principle" has a very narrow scope. The principle is not intended to eliminate or compensate for natural variations or for the contingencies of social life. Rather, it is designed to promote stability by allowing the least favored to gain functional advantages from the system of entitlements established by public institutions.[22] When prioritarians extend the concept to differences in health and fortune, they apply the concept to domains that Rawls does not intend it to permeate. In other words, paying careful attention to Rawls's own declarations about justice make it clear that broadening the scope of the "difference principle" into a general precept for the allocation of medical care would be a mistake. According to Rawls, the principles of justice that he offers, and "the difference principle" in particular, apply to the main public principles and policies that regulate social and economic inequalities. They are used to adjust the system of entitlements and earnings and to balance the familiar everyday standards and precepts. The difference principle holds, for example, for income and property taxation, for fiscal and economic policy. It applies to the announced system of public law and statutes, but not to particular transactions or distributions, not to the decisions of individuals and associations. Rather, the difference principle pertains to the institutional background against which these transactions and decisions take place.

Rawls concedes that the difference principle is not a constitutional essential,[23] and not a moral requirement.[24] For Rawls it is an important political consideration because of its relation to social cooperation and self-respect and, hence, social stability.[25] Particular medical care allocations are only required by political justice when citizens suffer a loss in fair equality of opportunity because of an accident or illness. Then society should aim at restoring people to the level of once again being "fully cooperating members of society."[26] But, the extent of even this political commitment to medical care is limited by costs and competing claims. Although, by footnote, he nods in Daniels's direction, Rawls explains the very limited application of the difference principle. For Rawls, restorative treatments "can be dealt with . . . when the prevalence and kinds of these misfortunes are known and the costs of treating them can be ascertained and balanced along with total government expenditure."[27] In other words, medical care does not have the priority for Rawls that it has for Daniels, and the provision of treatment is just one of several competing goods that a society should consider providing for its citizens.

Prioritarians who extend the difference principle to the health needs of all humans deviate from Rawls in several significant respects.

- For Rawls the difference principle is useful in the political domain; it is not a broadly applicable moral principle.
- For Rawls, it is primarily applicable to political arrangements and it is relevant to health care only to the extent that treatment can maintain or restore citizens as functioning members.
- For Rawls, the principle applies only to those who have the moral powers that enable them to cooperate in political arrangements: Those who use it as a broad moral principle tend to apply it to the worst off, but some such individuals are not, and could never be, citizens.
- Stability justifies the difference principle in Rawls's framework.
- Extending the principle, beyond the narrow parameters that Rawls sets, requires additional justification.

Those who want to apply the difference principle to additional domains or additional subjects, or to the provision of a broader array of medical services cannot claim to build on Rawls. They must provide justifications of their own for each extension.

JUSTICE IN ALLOCATIONS FOR TERRORISM, BIOLOGICAL WARFARE, AND PUBLIC HEALTH

The theoretical issues discussed above suggest two conclusions. First, a broad consensus supports the policies that were instituted in the wake of the terrorist attacks of September 11th and the anthrax attacks that followed shortly. Second, the leading candidates for a conception of justice, namely fair equality of opportunity and the difference principle, do not explain the positions that we accept on public health. The glaring incongruity between the post-September 11th policy consensus on the one hand, and these lauded principles of justice on the other, suggests that there is a mistake in our search for a ruling principle of justice. It also suggests an alternative for looking at the problem of justice.

It may seem surprising, but Rawls provides us with a solution to this dilemma. In *Political Liberalism* he explains that the aim of the difference principle is the maintenance of stability in society. In other words, a reason that reasonable people would accept supports the principle and its priority in the regulation of political arrangements. Large economic and social disparities destabilize a society, so because the significant advantages of communal life could not be maintained without prioritization for economic and social rewards, prioritization is a reasonable means for addressing the inevitable disparities in wealth and social status that develop in society. This insight implies that when stability is not threatened by a policy, prioritization may not be required or even justified. Different reasons could support different principles and different rankings of considerations in different situations. There is no obvious reason to presume that a single principle defines justice. Some reasons may be more defensible and more broadly accepted than others in different situations.

Triage may be the appropriate policy for large-scale emergency situations. The justification for triage is that adherence to a triage policy is most likely to *avoid the worst outcome* and to save the greatest number of lives. Reasonable people would want to survive a disaster and they would want their loved ones to survive. Foregoing treatment for those who are least likely to survive so as to provide the best chance of survival to the most people, therefore, yields the result that everyone wants most. In contemplating disaster scenarios and designing emergency response plans, it is hard to imagine any reasonable person objecting to triage and withholding their endorsement from the policy. So long as the same criteria for treatment are applied to everyone, the loved ones of those from whom treatment is withheld should not complain of injustice. Because their endorsement of triage policies can be legitimately presumed, a triage allocation of emergency services is not likely to undermine social stability.

Disaster preparedness requires the allocation of communal resources for research, training, and equipment. These are resources that could, instead, be devoted to improving the lot of those who are less advantaged than the rest. Yet, policies to allocate resources so that health professionals and other emergency service agencies can prepare themselves to respond to the emergencies that may occur are justified because, if a disaster should ever befall us, the ability to respond efficiently would be of vital importance. The good that can be had from preparation, would not be available without the prior contribution from a common pool. Hence, it would be reasonable for a society to provide some resources for disaster preparedness to increase the chance for a good outcome and to minimize the chance for the worst outcome (i.e., *maximin*).

In the face of a credible risk of biological warfare, mandatory inoculation against a contagious disease like smallpox is an appropriate policy when a reasonably safe and effective vaccine is available. Reasonable people would endorse such required inoculation because it provides protection from the disease, that is, it *provides a public good* that everyone values. Everyone should, therefore, bear a fair share of the burden of safety. Those who refuse to comply would be free-riders, treating others unjustly by taking advantage of their good will and sense of communal responsibility. Other public health measures, such as those that required the safe disposition of corpses after September 11th, are similarly justified by the public good of protection against disease that they provide and by the *anti free-rider principle* that would prohibit unsafe practices. And when it comes to actually dispensing the smallpox vaccine, because the relative differences between individuals may not be significant enough to be taken into account, a distribution scheme based on *equality*, such as first come first serve, may be required.

Furthermore, with respect to public health measures like vaccination, there may be good reasons for allowing a few to be exempt from inoculation. Those who are especially vulnerable to the inherent dangers of immunization, for example, those with impaired immune systems, would bear more than the typical burden of being vaccinated. If everyone else in the society received the vaccination, exempting those few who would otherwise bear an *undue burden*, would not increase the risk for others. In such cases, neither the implementation of the rule nor the countenance of legitimate exceptions would undermine social stability, so priority for the worse off would not be a relevant consideration. The difference in vaccination and exemption would not turn on some person or group being worse off than another. The difference in treatment would reflect the disproportionate burden on those who are immuno-compromised and the lack of any other justification for imposing that burden.

The public health concern about air quality immediately after the September 11th attacks reflects three slightly different principles. Clean air, clean water, and sewage treatment are the kinds of public goods that everyone needs constantly. Their *vital and constant importance to everyone's well-being* is a justification for policies to provide and protect them. In many environments they are also the kinds of benefits that no one can have unless everyone has them and making them available or unavailable at all makes them available or unavailable to everyone in the society. In many environments, these are also services that can be provided with most *efficiency* by providing them for everyone. Public health measures to assess air quality after the collapse and burning of the World Trade Center and the subsequent measures to protect the water supply from terrorist attacks are justified by both of these reasons.

Another important consideration is also relevant to the endorsement of public health interventions that provide for everyone's vital and constant needs. Such interventions are likely to make the greatest difference in health and well-being for the economically and socially least advantaged. The well-to-do could leave town for the clean air of the country or simply purchase gas masks to protect themselves from air pollution. They would also have the wherewithal to purchase bottled water, to dig private wells, and to install private sewage systems. The well-to-do would be better off with the general availability of clean air, clean water and sewage treatment. Yet, the underlying interrelation between poverty and health and the consequent disparity between the well-to-do and the poor with respect to health status and life expectancy,[28, 29, 30] suggest that the economically and socially disadvantaged would enjoy an even greater benefit from policies that made these benefits generally available. Furthermore, the continuous lack for some of such basic goods as clean air, clean water, and sewage treatment, when others enjoyed them as private resources, could promote social instability. The *difference principle* is, therefore, an additional reason for adopting public health measures to provide these services. It justifies the same policies that would be supported by the vital importance of the services and the fact that such services are most feasibly supplied to all at once. This example, therefore, illustrates how different appropriate principles can be just and converge in support of policies.

CONCLUSION

To the extent that policy domains covered by different principles can be legitimately distinguished, a variety of appropriate and compelling principles can express the complex and varied considerations that make different policies just. In other words, we should avoid the allure of a single simple ideal conception of justice. The just allocation of medical resources is and should be governed by a variety of considerations that reasonable people endorse for their saliency. Several principles have a legitimate place in public health allocation. They include: the *anti free-rider principle, avoid undue burdens, avoid the worst outcome,* the *difference principle, efficiency, equality, maximin, provide public goods,* the *vital and constant importance to well-being.*

To the extent that the scarcity of resources makes it impossible to fulfill all of the legitimate claims for society's allocation of resources, some principle(s) will have to be sacrificed and some projects that are supported by compelling reasons will have to be scaled down from an ideal level, delayed, or abandoned. When these hard choices have to be made, they too should be made for good reasons that reasonable people would support. In making difficult choices about the ranking of projects and priorities and the design of policies, different considerations will have different importance in different kinds of situations. There is no obvious reason to presume that one priority will always trump the others. When the priority of a principle reflects the endorsement of an overlapping consensus of reasonable people, the justice of the policy is clear. When people rank the competing considerations differently, a significant consensus on the principles that are irrelevant may emerge and that consensus can serve as the basis for just policy. To the extent that flexibility can be supported by the available resources, policies should show tolerance for different priorities.

As a general caution, however, those who make public health policy need to be alert to the kinds of illegitimate considerations that can distort and pervert any policy. Common psychological tendencies can interfere with judgment. For example, human psychology inclines people to exaggerate the impact of a loss and also inclines people to under-appreciate the value of future goods. This common inclination to inflate the importance of risk aversion seems to have contributed to the tremendous amount of resources invested in responses to the anthrax attacks and the charge into risky vaccination policies in the face of a conceivable but remote and unlikely small-pox threat. Politics is another factor that may have contributed to the huge investment in anthrax decontamination at sites where the likelihood of danger was almost non-existent. The desire to curry favor or to shore up votes does not promote reasonable public health policy. Prejudice, stereotyping, the desire to do something, the pressing needs made vivid by individual cases, lack of insight, and lack of foresight are other common psychological inclinations that can distort judgment and lead to unjust public health policies.

In sum, it is difficult to achieve justice in public health policy because there is neither a single ideal governing principle nor a simple formula for success. A variety of considerations can legitimately support good policy. A variety of cloaked and disguised hazards can intrude on the policy making process and, in the name of justice, lead to indefensible results. For public health policies to be just, the reasons behind them must be the ones that reasonable people would find most compelling and most appropriate. Policies must reflect the choices that reasonable people would make and the priorities that reasonable people find most pressing.

NOTES

[1] Rosamond Rhodes, "Understanding the Trusted Doctor and Constructing a Theory of Bioethics. *Theoretical Medicine and Bioethics*, 2001; 22(6): 493-504. I have argued generally that physicians have a role-related responsibility to avoid making judgments about patients worthiness and that they must treat all patients similarly based on medical considerations.

[2] Although a question for biomedical research may disproportionately affect a relatively disadvantaged population (e.g., the effect of lead paint on child development), the study findings and the public health policies that reflect the research findings, will have implications for all of those who have been or who may be affected.

[3] John Rawls, *A Theory of Justice*. (Cambridge, Mass.: Harvard University Press, 1971).

[4] John Rawls, *Political Liberalism*. (New York, NY: Columbia University Press, 1993).

[5] Rawls, 1993, p.6.

⁶ Rawls, 1993, p.4.

⁷ Rawls, 1993, p.184.

⁸ Rawls, 1993, p. 326.

⁹ Rawls, 1993, pp. 228-29.

¹⁰ Rawls, 1993, p.6.

¹¹ Norman Daniels, "Justice, Health, and Health Care" in *Medicine and Social Justice: Essays on the Distribution of Health Care*, Rhodes, Battin and Silvers, editors. (New York, NY: Oxford University Press, 2002), pp. 6-23.

¹² Daniels, 2002, p. 8.

¹³ Julian Savulescu, "Justice and Healthcare: The Right to a Decent Minimum, Not Equality of Opportunity," *American Journal of Bioethics*, 1.2 (Spring 2001) online.

¹⁴ Ronald M. Green, "Access to Healthcare: Going Beyond Fair Equality of Opportunity." *American Journal of Bioethics*, 1.2 (Spring 2001): 22-23.

¹⁵ Dan W. Brock, "Priority to the Worse Off in Health-Care Resource Prioritization," in *Medicine and Social Justice: Essays on the Distribution of Health Care*, Rhodes, Battin and Silvers, editors. (New York, NY: Oxford University Press, 2002), pp. 362-372.

_____ "Aggregating costs and benefits," *Philosophy and Phenomenological Research* , 58 (1998):963-68.

¹⁶ F.M. Kamm, "Whether to Discontinue *Non*futile Use of a Scarce Resource," in *Medicine and Social Justice: Essays on the Distribution of Health Care*, Rhodes, Battin and Silvers, editors. (New York, NY: Oxford University Press, 2002), pp. 373-389.

_____ *Morality, Mortality Vol I: Death and Who to Save From It*. (New York, NY: Oxford University Press, 1993).

¹⁷ David Wasserman, "Aggregation and the Moral Relevance of Context in Health-Care Decision Making," in *Medicine and Social Justice: Essays on the Distribution of Health Care*, Rhodes, Battin and Silvers, editors. (New York, NY: Oxford University Press, 2002), pp. 65-77.

¹⁸ Rawls, 1993, p.328

¹⁹ Rawls, 1993, 284 & xvii.

²⁰ Rawls, 1993, xvii.

²¹ Bernard Gert, Charles Culver, and K. Danner Clouser, *Bioethics: A Return to Fundamentals*, (New York, NY: Oxford University Press, 1997).

²² Rawls, 1993, p. 283.

²³ Rawls, 1993, pp. 228-29.

²⁴ Rawls, 1993, pp.236-37, fn. 23.

²⁵ Rawls, 1993, p. 318.

²⁶ Rawls, 1993, p. 184.

²⁷ Rawls, 1993, p. 184.

²⁸ Daniels, 2002.

[29] Patricia Smith, "Justice, Health, and the Price of Poverty." in *Medicine and Social Justice: Essays on the Distribution of Health Care*, Rhodes, Battin and Silvers, editors. (New York, NY: Oxford University Press, 2002), pp. 301-318.

[30] Mark Sheehan and Peter Sheehan, "Justice and the Social Reality of Health: The Case of Australia,"in *Medicine and Social Justice: Essays on the Distribution of Health Care*, Rhodes, Battin and Silvers, editors. (New York, NY: Oxford University Press, 2002), pp. 169-182.

JACQUELYN ANN K. KEGLEY

A NEW BIOETHICS FRAMEWORK FOR FACILITATING BETTER DECISION-MAKING ABOUT GENETIC INFORMATION

ABSTRACT: It is argued that a new philosophical framework is needed to more adequately address medical decision-making about genetic disease. Persons need to be reconceived as complex relational networks and disease and illness redefined to better reflect their profound impact on all aspects of a person and their life. Decision-making needs to be viewed as involving more than an isolated individual right-bearer and autonomy and responsibility, individual and community need to be more carefully balanced. A modified human rights model is discussed and illustrated through discussions of genetic testing. Such a model could also reorient public health decision-making.

KEYWORDS: Asymptomatic patient; multimorbidity; relational networks; multiple deficits; narrative beings; subjective substantive disclosure rule; dialogue.

Notions of paradigm shifts and Copernican revolutions perhaps, like all faddish ideas, have outlived their usefulness. Yet, on past occasions, I have suggested that scientific breakthroughs in molecular genetics and developments in genetic technology demand new modes of thought in philosophy, ethics, law and public policy.[1] It has also been my contention that elements of such new frameworks are not so new, but rather are to be rediscovered in the philosophical work of the American philosopher, Josiah Royce.[2] Others have found their inspiration for new approaches, such a Clinical Pragmatics and Pragmatic Bioethics, in the thought and writings of John Dewey.[3] In what follows I will briefly establish that genetic knowledge and genetic disease have unique characteristics that require reformulation of our usual medical modes of thought as well as our attempts to deal with moral problems related to genetic issues. Secondly, I will suggest the outline of a new philosophical framework which I believe to be better suited to facilitating more adequate decision making concerning the use of genetic information. Such a framework might also help "redefine" public health.

What, then, do I believe needs to be changed in light of new genetic information? First, the concepts of "gene" "disease" and "patient" need to be de-individualized. A gene is not a solitary causal mechanism. The immense complexity of the human genome and the highly interdependent and interactive nature of genetic mechanisms

91

M. Boylan (ed.), Public Health Policy and Ethics, 91-101.
© *2004 Kluwer Academic Publishers. Printed in the Netherlands.*

needs to be recognized and genetic determinism and genetic essentialism need to be seen as faulty explanatory schemes. Few diseases are caused only by genetic factors and persons do not equal the sum of their genes. In fact, a new understanding of medicine, "molecular medicine," is based on the premise that disease and health must be understood "in terms of the interaction of genes and environment (everything other than genes)."[4]   Further, genetic disease is not necessarily about a single patient. Rather genetic problems may be shared with family members and future offspring and genetic treatments and information may have significant benefits and liabilities for others. Again, another premise of the new molecular medicine is that "fiduciary obligations are owed by physicians to populations of patients who are bound to each other by various social, moral, and genetic bonds, as well as to individual patients."[5]

Secondly, genetic disease impacts on both individual and group identity. Those who discover they have a genetic problem often take it quite personally as a blemish of their very person. Defective genes mean defective people. This close identification of gene with person and personal worth is in fact reinforced by the genetic essentialism often espoused by scientists and journalists, namely, that genes make us who we are.[6] This identification of genes with self not only promotes a strong feeling of guilt among those identified with a genetic problem, but also brings to prominence concerns about autonomy and privacy. Genetics matters have much to do with one's self, one's future and with decisions regarding that future. Genetic information is crucial to personhood and thus decisional privacy becomes an important issue. The label of a genetic problem can also stigmatize the individual and subject them to forms of discrimination in employment, in insurance, and in the eyes of social others. And, it can do the same for others with whom the person has a biological or social connection. Genes connect with race, gender, and ethnicity and thus attempts to screen for certain genetic diseases such as Tay Sachs and Sickle Cell Anemia, both of which are tied to certain racial and ethnic groups, have led to charges of eugenics and discrimination. These charges become even more plausible when prevention techniques, e.g. screening, are not preventing the disease, which is untreatable, but preventing the birth of individuals with the disease.

In addition, genetic disease is also a family matter, and thus familial and cross-generational decisions may require genetic information that has been given to one family member, the patient. BRCA 1 and BRAC 2 are inherited mutations which run in families and which can lead to breast or cervical cancer. If a mother or sister discovers she is the carrier of the BRAC gene, what obligation is there is reveal this information to their female relatives who may be at risk? This, of course, raises issues of confidentiality, especially when the information is needed to prevent serious harm to another. In mental health law, there is the Tarasoff rule that concerns breach of confidentiality when a clear and imminent danger exists for an identifiable third party. Maybe such a law will eventually be needed to deal with genetic information. The use of genetic information does seem to provoke questions about decisional privacy and confidentiality and call for a new look at traditional medical, legal and ethical policies.

A final complication with genetic disease is the phenomenon of the "asymptomatic patient," a person who has a genetic defect, but who does *not* now or will <u>not</u> ever have the full-blown disease manifestation. In this case social contexts and actions may impact heavily on these persons and their autonomy and privacy may be under serious attack. Thus, the symptomatic person may be given a social status of "sick," which can lead to self-image problems, loss of employment, and, as in cases of sickle cell trait (which is not the same as sickle cell anemia), exclusion from military service as a pilot or from exercising other choices.

The asymptomatic patient phenomenon leads to three characteristics of genetic health information that seems to distinguish it from conventional health information and diagnostics. First, genetic disease can be predicted long before it appears symptomatically. Sociological consequences of such a diagnosis were discussed above, but what about the psychological consequences for an individual who discovers that a serious and incurable disease will befall them in the future? Second, genetic dispositions to diseases and susceptibilities to exacerbating environmental circumstances can also be predicted. Again, there are serious psychological and sociological consequences of such predictions, but there are also serious implications for definitions of health and disease. In light of this phenomenon, German philosopher Kurt Bayertz projects the idea of a "universal presymptomatic multimorbidity" and argues that "everybody will be 'diseased' – but in a hidden and ambiguous sense."[7] Thirdly, genetic diagnoses permit the attribution of particular risks to particular individuals and this in turn seems to demand preventive measures and a new sense of responsibility. It might not be inappropriate to place social, legal, financial and ethical responsibility on those guilty of omitting to take preventive measures. A new understanding of individual responsibility seems to be needed.

A New Framework

In light of the issues raised by the new genetics, I wish to suggest the outline of a new philosophical way of looking at these concerns. First, "persons," whether patients or other decision-makers, must be seen as complex relational networks, composed of a variety of public physical aspects— material, neural, genetic, behavioral, social, cultural, political, economic— and a mix of private, inner aspects — sensual, emotional, mental, and intentional. Further, all these aspects must be seen as interacting with and influencing each other in complex and multiple ways. Individual persons are holistic and should not be reduced to any one of their aspects, whether genetic or endowed rights. [8] Further, "persons" must be seen as in process, creatures of time, capable of stagnation, development, and growth and able to take advantage of opportunities of the future. Persons can and should be able to think in terms of both short term and long-term goals and events. This is especially important for genetics where one deals with probabilities, new horizons of information and a future of new technologies and treatments. All of us know the uncertainties of prognosis for many genetic-based disorders, e.g. Down's Syndrome, Spinal Bifida, etc. One should not give up too easily on the prospects for the flourishing of various afflicted individuals, especially if there is strong support from family and others. And, one should not be too quick with diagnosis and treatment, but rather should be open to discussion of various possibilities and suggestions from a variety of areas of expertise. Indeed, diversity should be valued since it is often the key to survivability in the genetic scheme of things.

Secondly, "illness" or "disease" must *not* be seen as located solely in a physically malfunctioning body, but rather be viewed as impacting the persons at a multiplicity of levels. [9] In addition to "deficit," perhaps "multiple deficits" in various systems of the bodily organism is more appropriate. Further illness, whether genetically based or not, impacts psychological functioning. It often is a "fundamental shattering of everyday personal assumptions" and a blow to one's total personhood and life plans as an individual.[10] Illness is also clearly social in nature; it involves loss in terms of social functioning and an inability to meet individual and societal expectations. Illness often involves failure of obligations and disruptions of relationships. My own analysis of disease and illness is in concert with that of Micah Hester who sees disease and injury as implying social disruption, and breakdown of social vitality and the "everyday" and "taken-for-granted" quality of our experiences.[11] In dealing with genetic disorders and genetic information all of these aspects of "disease" and "illness" need to be taken into account. What kind of dysfunction does the genetic information indicate? What kind of "abnormality," is it? Is it biological, psychological, sociological, cultural, or even a matter of individual perception? Or, is it a combination of all of these factors? Further, given this genetic information, what type of "flourishing" still might be possible for the affected individual, given that individual social structure and support systems? And one must assume persons can reformulate their patterns of living and forms of vital functioning. Hester describes two cases of such reformulation in his discussion of disease and health. One of these is the artist who lost his sense of color in an accident, but who, with help, found a new world of subdued light, the world of night. He was restored to health, namely, "living healthily," described by Hester as "actively pursuing goals and developing your*self* by accounting for both your on-going story line and the current demands of the environment."[12]

Thirdly, in the new framework, "persons" must be seen as relational in a fundamental sense, namely, as a being-with-others. Individuals are usually engaged in life in a mutual journey with others with whom they share values, goals, goods, liabilities and risks. These relationships can be biological and/or social and "kinship," of either kind or both is a powerful aspect of human life. Thus, anthropologist, Robin Fox, writes: "The relationships to ancestors and kin have been the key relationships in social structure; they have been the pivots on which most interaction and most claims and obligations, most loyalties and sentiments turned." [13] A closely connected relational aspect of human individuals is their nature as "narrative beings," i.e. those who construct their own stories, interpreting and interconnecting past, present and future.[14] These narratives usually, if not always, include others, especially those with whom the individual has some genetic connection. Our narrative connections, which include the genetic ones, give cohesiveness and quality to our individual lives and make us feel situated and recognized as individuals. Our kinships, both social and biological, are a basic dimension of our human identity and of our flourishing. As such, they must be taken into account in using genetic information and in decision-making that involves genetic information.

Fourthly, in the new framework, a new understanding of autonomy and related concepts like confidentiality and privacy must be forged. In the traditional medical and legal context, the principles of autonomy, privacy and confidentiality have centered emphasis on the right of individual persons to make personal decisions without interference. In this light, genetic counseling has been non-directive, i.e. the counselor functions as a neutral purveyor of information. Medicine has generally been practiced within the culture of individual rights with emphasis on informed consent, on the right to refuse treatment, the right to decisional privacy such as guaranteed in procreative liberty, and on the right to control medical information about themselves. However, individual rights are never without limits and, as we have already seen, genetics raises important questions about where individual rights end and where responsibilities to others such as one's family or the larger society begin.

Much ethical analyses emphasize the self-chooser, usually in isolation from others, and as a bearer of rights who must be protected from the interference of others. These emphases are important and should not be disregarded, but they should be placed in context. Part of that context is the recognition that ascriptions of rights are, in a most fundamental way, ascriptions of others and dependent upon others. The recognition of rights is communal and the accompanying moral duties and obligations are grounded in complex social practices such as parenting and the practice of medicine. It may well be useful to view rights as fallbacks, i.e., as efforts to obtain something that could not be better assured by ties of affection and loyalty or by moral duties deeply ingrained in our complex social practices such as parenting and health caring.[15] As for self-choice, the notion of an isolated self-chooser is belied by the recognition of human individuals as essentially "beings-with- others. And, as Charles Taylor has argued in *The Ethics of Authenticity*, if choice is its own rationale, then morality is trivialized.[16]

In light of the tensions between individual autonomy and responsibility to others, some have suggested that a public health model should be considered for genetics and genetic disease. Such a model encourages prevention of disease including certain mandatory medical interventions such as vaccination and seeks to warn persons of health risks through education. In some cases, such as smoking in the U.S., the law has stepped in to protect persons, both individuals who are engaging in the risk and others, from health risks. In fact, the public health model has been used in genetics in the United States to argue for mandatory genetic screening for Tay Sachs and Sickle Cell Anemia. However, there are problems with applying the public health model to genetic issues, some of which I have discussed on other occasions. I believe the public health model is not appropriate for dealing with genetic diseases, but rather a modified human rights model needs to be applied which gives much more attention to the limits on individual rights in light of communal and familial ties and common vulnerabilities. However, it also might be fruitful to redefine the traditional public health model in terms of the philosophical framework proposed in this essay.

To give some indication of a modified human rights model for genetic decision-making, I will briefly discuss how genetic testing might be viewed in such a model. Testing should be done in the context of autonomy and informed consent, but with regard to the dangers that might be posed for others. All testing must be premised on providing adequate information before testing is undertaken. The standard of "informed consent" that should be applied is that of the *subjective substantial disclosure rule,* that is, the information provided is that which would be material or important to the decision of this particular patient in this circumstance. A key with this rule is to ask: Could this information change the decision of this particular person in this particular circumstance?" Such a rule requires a substantial degree of knowledge about the patient, her context, and what is important to her and, I would argue, to her significant others. People are influenced in their decisions by the views of significant others and they are generally concerned about the impact of their decisions on the lives and health of those they know and care about. Further, giving such information should be considered a priority in operating under the subjective substantive disclosure rule.

In the context of genetic screening, this implies that substantial genetic counseling and education should take place before the screening is undertaken. Generally there is time for careful counseling since most genetic tests do not concern immediate and serious risk factors. The patient needs to be informed about the severity, potential variability, and treatability of the disorder being tested for. Those providing the counseling must be sensitive to the fact that notions of risk are viewed differently by individuals and often vary considerably because of social contexts. Disease, disorders and risk are viewed differently by different ethnic, racial, cultural groups and even by the genders.

Adult patients should be informed of the possible implications of the genetic information for others in their families and encouraged to consider sharing information with them. The choice should be left to them unless there is a clear and imminent danger to identifiable others. It is my judgment that breach of confidentiality may be warranted if there is clear and imminent danger to another or other identifiable persons and the patient, after reasonable discussion, refuses to allow information to be revealed. The conditions under which confidentiality might be breached should be discussed with the patient at the beginning of the counseling relationship.

Prenatal screening and screening which concerns decisions about children or potential children is a more difficult matter. Mandatory newborn screening (e.g. for PKU and hypothyroidism) is done with the goal of providing early treatment that can make a dramatic difference in the child's well-being by preventing mental retardation. Such screening is done under *parens patriae* to prevent substantial, imminent harm to the children. It is my judgment, that this operating rule for newborn screening is a good one. However, this type of rule would raise serious questions about screening newborns for untreatable disorders or carrier status. These children may suffer stigmatization or rejection from parents and might become uninsurable, unemployable and unmarriageable. Further, children may be provided with information that, at age of consent, they would rather not have. In other words, information can cause serious harm to these children and thus it is not always desirable to seek.

Even if the screening is for treatable diseases, similar questions can be raised. An additional element is the range of predictability of the tests. Thus, for example, studies with Cystic Fibrosis showed that "only 6.1 percent of infants with positive first tests were ultimately found to have cystic fibrosis on sweat chloride testing." Yet, parents continue to consider their children "sick" and to have concerns for their health. Such parental reactions could negatively impact on the ultimate health and life prospects of these children. Such children are, in fact, at high risk for a serious disorder of psychosocial development known as the "vulnerable child syndrome." Another important question to ask about screening for treatable diseases is whether the proper support mechanisms for treatment are available. What if there is no assistance for the expensive diet regime needed for PKU? Many children with sickle cell anemia, for example, do not get their necessary penicillin prophylaxis. If one insists on tests and on providing information, who has the obligation to follow through on providing treatment and the fulfillment of the goal to prevent disease or eliminate or relieve its symptoms?

The necessity of subjective substantive information is even more evident in prenatal screening where the impact of the decisions is of great import. This is a situation that can too easily be seen as one of too few options or of binary choices. Every effort must be made to explore the full range of possibilities for treatment, support, etc. The question of severity of the disorder must be addressed thoroughly and in the context of the individual's values and beliefs and social, cultural, ethnic context. One needs to be aware that a woman's religious background is often a key factor in the decision-making and that women with lower educational attainment and income often show more willingness to have an affected child. These factors, however, should be assessed in the full context of social, psychological, and economic support. One also needs to be aware that the very testing process can interrupt the formation of maternal-infant bonding, a phenomenon known as "tentative pregnancy." In the context of autonomy, and decisional privacy exemplified in procreational liberty and the right of the woman to make decisions in this situation, it is imperative I believe that health care providers not be seen as neutral dispensers of genetic information. There must be careful and interactive conversation about the meaning of genetic information, about the accuracy and predictability of such information and it needs to be explained and understood in the 'lived context" of those who must make the decisions and then "live with them." There clearly also needs to be a recognition that individuals "alone" facing difficult decisions involving complex information may act too hesitantly or too impulsively. Their decisions, though impulsive and uninformed, may have profound consequences for others. Persons, facing these decisions, need all the help they can get from professionals, and significant others and this should be encouraged. Procreative liberty exercised individually can be dangerous. And, I agree fully with the judgment of Glenn McGee that the so-called non-directive counseling which requires the physician and genetic counselor to dispense neutral information about disease without making evaluative commentary is just as dangerous as paternalistic medicine.

Indeed, our new framework should see decision-making, whether in the genetic realm or another, as necessarily involving more than an individual. It would be seen in terms of individuals in dialogue, a dialogue that could include the primary individual, significant others, some (such a potential offspring) represented through others, health care professionals, counselors, and others with significant interest in the outcome. The relevant individuals to be included in the discussion could be identified in terms of the following set of circumstances, as suggested by Rosamond Rhodes in a recent article on rights and responsibilities in the face of genetic knowledge,[17] namely: (1) Family relationships, defined both biologically and socially; (2) Social relationships; (3) The history of the relationships; and (4) The particulars of the relationship and situation.

The dialogue should be based upon at least three premises. First, there should be efforts to broaden the dialogue and decision-making context with at least three goals in mind: (1) to include those others likely to be significantly impacted by the decision in question or whose values, interests and goals are crucial aspects of the individual decision-makers own life story; (2) to gather more relevant information for the decision to be made including the facts particular to the case and context; and (3) to stress a democratic model of moral problem solving which recognizes that moral problems in clinical practice cannot and should not be solved by expert judgment alone. [18] This third goal is one stressed by "Clinical Pragmatism", but it also follows from our own view about persons as both relational and narrative. And it is my judgment that Clinical Pragmatism needs to refine its position to make clearer the role of ethical principles and of expertise. A good beginning of defining the role of expertise in a more democratic model of moral decision-making is the efforts of the task force, appointed jointly by the Society for Health and Human Values and The Society for Bioethics Consultation to develop a set of standards for Bioethics Consultation. In the Task Force's recent discussion draft they argue that the *"qualified facilitation model"* is more appropriate for health care ethics consultation in U.S. society. They write: "This is because a *qualified* facilitation role taken by the bioethics consultant is consistent with the fact of pluralism and the political rights of individuals to live by their own moral values." [19]

Further, another emphasis in the dialogue should be about respect for each voice and individual, while seeking mutually agreed upon choices in the situation at hand. There would be a moral commitment to mutual interpretation and understanding so that each voice would be "genuinely heard." To facilitate this kind of interpretation and understanding is, of course, a special skill, and the Task Force on Standards for Bioethics Consultation advocates "process and interpersonal skills" as a second core skill for anyone who undertakes the task of "facilitating decision-making and "helping patients, families, surrogates, health providers, or other involved parties address uncertainty or conflict regarding value-laden issues that emerge in health care."

Finally, the dialogue model of decision-making would be governed by the principle of "loyalty to loyalty," namely the principle of ever-broadening community in the sense of fostering conditions for the increased autonomy and flourishing of individuals and groups.[20] This means the need to take account of the "full" genetic, environmental, social, organizational and individual context of the situation and of the individuals most critically affected by the decisions. This is, of course, an ideal that gives us goals to work toward in dealing with the complexity of using genetic information in a "fitting" manner.

The new framework sketched here clearly needs to be developed in much more detail. However, I have argued that the nature of genetic information and genetic disease creates a serious problematic for our traditional individualistically based frameworks in medicine. Public health models have been less individualistic and more "communal and socially based. However, they tend to neglect the balance of individual and community. Public health models should also be re-examined in light of my proposed framework. The individualistically framed medical model, especially as manifested in "genetic essentialism," poses a very real threat to individuals who suffer from "genetic disability," and given that such traits are carried by us all, it threatens us all. This means that together we must forge a new paradigm to guide the judicious use of genetic information. It should be one that fosters human autonomy in the context of community and that respects individuals both as unique and yet dependent upon and formed by relational context. Only then might we be able to adequately deal with genetic disease, genetic health and human persons as genetic, but who are also so much more than those genes.

NOTES

[1] Jacquelyn Ann K. Kegley. 1998. "A New Framework for the Use of Genetic Information," in *Genetic Information: Acquisition, Access, and Control*. Edited by A. Thompson & R. Chadwick. (New York: Plenum Press); Jacquelyn Ann Kegley. _____. "Using Genetic Information: A Radical Problematic for an Individualistic Framework." *Medicine and Law* 15.4 (December, 1996); and Jacquelyn Ann K. Kegley, "Genetic Information and Genetic Essentialism: Will We Betray Science, the Individual and the Community?" in *Genetic Knowledge: Human Values and Responsibility*. Edited by J. Kegley. (New York: Plenum Press, 1998).

[2] Jacquelyn Ann K. Kegley, Gen*uine Individuals and Genuine Communities: A Roycean Public Philosophy*. (Nashville, Tennessee: Vanderbilt University Press, 1997).

[3] Joseph J. Fins, Matthew D. Bacchetta, and Franklin G. Miller. "Clinical Pragmatism: A Method of Moral

Problem Solving." *Kennedy Institute of Ethics Journal.* 7.2. (June, 1997): 129-146; and Glenn McGee. *The Perfect Baby: A Pragmatic Approach to Genetics.* (Boulder: Rowman & Littlefield, 1997).

[4] Laurence B. McCullough. "Molecular Medicine, Managed Care, and Moral Responsibilities of Patients and Physicians," *Journal of Medicine and Philosophy.* 23.1 (1998): 3.

[5] McCullough, 1998, p. 7.

[6] James Watson, Co-Founder of the DNA double helix has proclaimed that it is our DNA which makes us human. He has also said "Our fate is in our genes." Quoted in Leon Jaroff, "The Gene Hunt." *Time.* 20.(March 19, 1989): 62-67. For a close examination of the prominence of genetic essentialism in American culture see: Dorothy Nelkin and M. Susan Lindee, *The DNA Mystique: The Gene as a Cultural Icon.* (New York: W.H. Freeman, 1995).

[7] Kurt Bayert, "What's Special about Molecular Genetic Diagnostics?" *Journal of Medicine and Philosophy.* 23.3 (1998): 250.

[8] Jacquelyn Ann K. Kegley, "Peirce and Royce on Person." in Herman Parret Ed. *Peirce and Value Theory.* (Amsterdam and Philadelphia: John Benjamin Publishing, 1994), pp. 17-26.

[9] For a discussion of "illness" viewed in this manner see: Jacquelyn Ann K. Kegley, *Genuine Individuals and Genuine Communities.* (Nashville, Tennessee: Vanderbilt University Press, 1997).

[10] Kegley. *Ibid.*

[11] D. Micah Hester, "The Place of Community in Medical Encounters." *The Journal of Medicine and Philosophy.* 32.4. (1998): 369-83.

[12] Hester. p. 374.

[13] Robin Fox, *Kinship and Marriage: An Anthropological Perspective.* (Baltimore, Maryland: Penguin Books, 1967), p. 86.

[14] James Lindemann Nelson. "Genetic Narratives: Biology, Stories, and Definition of Family. In Cynthia B. Cohen. *New Ways of Making Babies.* (Bloomington, Indiana: Indiana University Press, 1997), pp. 51-69. See also: Kegley, 1997. *Genuine Individuals and Genuine Communities*, chapter 3.

[15] See: Thomas Murray, "New Reproductive Technologies and the Family." in Cynthia B. Cohen. *New Ways of Making Babies.* (Bloomington, Indiana: Indiana University Press, 1997), pp.. 51-69.

[16] Charles Taylor, *The Ethics of Authenticity.* (Cambridge, Massachusetts: Harvard University Press, 1993).

[17] Rosamond Rhodes, "Genetic Links, Family Ties and Social Bonds: Rights and Responsibilities in the Face of Genetic Knowledge." *Journal of Medicine and Philosophy.* 23.1 (1998):10-30.

[18] See: Joseph J. Fins, Matthew D. Bacchett, and Franklin G. Miller, "Clinical Pragmatism: A New Method of Moral Problem Solving." *Kennedy Institute of Ethics Journal.* 7.2 (1997): 129-145.

See also: Rosemarie Tong, "The Promises and Perils of Pragmatism: Commentary on Fins, Bacchetta, and Miller." *Kennedy Institute of Ethics Journal.* 7.2 (1997): 147-152.

[19] "Discussion Draft of the SHHV-SBC Task Force on Standards for Bioethics Consultation" 7.

[20] For an extensive discussion of this notion of loyalty and the process of interpretation,

See: Jacquelyn Ann K. Kegley, 1997, *Genuine Individuals and Genuine Communities*, chapters 4 and 7.

MARY B. MAHOWALD

SANCTITY OF LIFE VS. QUALITY OF LIFE IN MATERNAL-FETAL SURGERY: PERSONAL AND PUBLIC PRIORITIES

ABSTRACT: Maternal-fetal surgery for repair of fetal spina bifida is an experimental treatment that requires research to document its therapeutic effectiveness. Such research cannot be ethically performed unless non-medical benefits are considered. Personal decisions are influenced by public priorities, and sanctity of life considerations mingle with quality of life considerations in those who seek the procedure. Critique of dichotomous interpretations of these standards is crucial to development of public policies regarding its availability. Once different sides of the apparent dichotomies are recognized as overlapping, covert discriminatory attitudes and practices toward people with disabilities may be uncovered and addressed.

KEY WORDS: maternal-fetal surgery, spina bifida, disabilities, quality of life, sanctity of life, equipoise, public-private distinction

Personal and public priorities with regard to prenatal testing and treatment have been morally problematic for many years. An obvious reason for this is ongoing debate about the moral status of human embryos or fetuses.[1] Depending on where one stands on that issue, arguments for reproductive liberty are more or less compelling. Other morally relevant factors are empirical: the risks, costs, and potential benefits of particular tests and treatment for pregnant women, their potential children, and the public at large. Among those affected by policies that address the topic are those who oppose it on grounds of "ableism," i.e., discrimination against people with disabilities; this argument is supported by the fact that prenatal testing and termination are usually undertaken to avoid the birth of a child who is disabled.

In this essay I focus on a type of maternal-fetal surgery intended to reduce the incidence or severity of disability in children. Although this treatment has been privately pursued and personally funded by an increasing number of women and couples, empirical research is necessary to determine whether it ought to be offered to the public at large, and covered for those who are unable to afford it. Whether such research can be ethically conducted has been questioned by those who favor the surgery as well as by those who oppose it. I argue that the research can be ethically per-

M. Boylan (ed.), Public Health Policy and Ethics, 103-118.
© *2004 Kluwer Academic Publishers. Printed in the Netherlands.*

formed, but only under certain conditions. With regard to women who seek and obtain the treatment, there is need to examine whether public attitudes and practices compromise the autonomy of their decisions. Public priorities, I argue, may obscure or distort the personal priorities of individuals, even to the individuals themselves. Support for this particular treatment suggests that quality of life considerations have priority over sanctity of life considerations in both domains. However, because quality of life considerations presuppose some respect for the sanctity of life, the two domains inevitably overlap.

MATERNAL-FETAL SURGERY FOR REPAIR OF MYELOMENINGOCELE

Myelomeningocele, a form of spina bifida, is a congenital anomaly diagnosed in about one of every 1000 pregnancies a year in the United States.[2] It is the principal non-lethal condition for which physicians have performed maternal-fetal surgery. Onset is triggered by failure of the primitive neural tube to close at about the fourth week of gestation. The main impairments that affected individuals suffer are loss of motor and bladder control; a significant minority also experience bowel incontinence and developmental delay. Currently, the standard of care for affected infants is surgery to close the lesion to prevent infection and limit neurological damage. Clinicians are legally and morally bound to meet this standard, defined by one law professor as "what a reasonably prudent physician would do in the same or similar circumstances."[3] According to Lyerly et al., the majority of maternal-fetal specialists do not consider in utero treatment of nonlethal conditions to be standard of care. Fifty-seven per cent support a moratorium on the procedures until a multicenter-controlled clinical trial demonstrates that surgery before birth is at least as therapeutically successful as postnatal treatment.[4]

When spina bifida is detected prenatally, the pregnant woman may terminate the pregnancy or proceed with gestation and prepare for her infant to be treated appropriately after birth. Misleadingly, the latter alternative has been characterized by one of its principal practitioners as "doing nothing."[5] He and his colleagues consider the treatment they have been offering prior to birth as a third alternative, one that they apparently encourage women and couples to pursue.

The prenatal interventions intended to correct spina bifida are performed between 21 and 30 weeks gestation through endoscopic or open surgery on the pregnant woman. Both methods have resulted in serious fetal and maternal complications.[6] With open surgery, incisions are made in the woman's abdomen and uterus to expose the fetus for neurosurgical repair, usually performed by a pediatric neurosurgeon. Following surgical closure of the neural tube, the fetus is returned to the uterus and all of the incisions are closed; the woman then continues gestation. Her obstetric care is provided by physicians other than those who perform the maternal-fetal surgery.

The assumption that underlies attempts to repair spina bifida in the fetus is that the major impairments associated with it may become more severe if the surgery is delayed. Unfortunately, although the prenatal procedure has thus far been performed on more than 150 women, there is no evidence that motor, cognitive, or bladder function have improved in the children to whom they subsequently gave birth. Endoscopic repair, which carries fewer risks for the pregnant woman, has been attempted less often, apparently because direct surgical repair is more likely to reduce impairments in the newborn.

Two clinical outcomes of maternal-fetal surgery for repair of spina bifida may be viewed as secondary benefits: postponement of the need for shunting, and reduction in hindbrain herniation.[7] However, no improvement in neurological outcome has been associated with either of these results. Moreover, intervening before birth introduces risks that would not otherwise occur for the potential child or for the woman. These include not only the usual risks of anesthesia and surgery, but also hemorrhage, chorioamnionitis, thrombosis, tocolytic-induced pulmonary edema, and uterine rupture. Classical cesarean section must be performed at delivery not only for this pregnancy but in subsequent pregnancies as well; other obstetric risks also apply to future pregnancies. In addition, preterm labor, premature rupture of membranes, and oligohydramnios are likely in 50% of women who undergo the procedure.[8]

The facility at which open surgery for fetal spina bifida has been performed more often than at any other institution is Vanderbilt University. Most of the potential patients who come to Vanderbilt in pursuit of the procedure are attracted to the institution through its website, which states that the surgeons who perform the procedure "are no longer using an experimental protocol."[9] In light of the risks of treatment and lack of evidence that its primary goals have been met, this statement is misleading and unjustified. During conferences at Vanderbilt and at the National Institutes of Health in 2000, the language on the website was questioned, and the Vanderbilt team was urged to acknowledge the experimental status of the procedure.[10] An apparent result of the conference convened by the National Institutes of Health was the agreement of practitioners from programs that have been offering maternal-fetal surgery during the past decade to develop a multi-center protocol designed to obtain evidence of the therapeutic effectiveness of their attempts to repair spina bifida during gestation. A five-year multi-center trial has now begun (March 2003) to determine whether surgery during gestation is more beneficial to the potential child than surgery after birth.[11] Because there are no potential medical benefits to the pregnant woman, the same criterion cannot be met in her regard.

Prior to his participation in the multi-center trial, Joseph Bruner, the obstetrician who launched and still directs the program at Vanderbilt, expressed doubt that ethical research on maternal-fetal surgery for spina bifida is possible. On grounds of his experience with potential patients, he said that they are generally so convinced of the effectiveness of the procedure that they would not agree to randomization that might deprive them of it.[12] In light of this reluctance, he questioned whether the ethical requirement of equipoise could be met. Bruner acknowledged that no improvement in mobility or continence had been demonstrated, and that further risks are introduced by performing the surgery before birth. Nonetheless, he apparently believes that postponement of shunting and reduction in hindbrain herniation justify the procedure.

In experimental research on human subjects, equipoise is defined as a state of being "equally poised in our beliefs between the benefits and disadvantages of a certain treatment modality."[13] Achievement of equipoise is generally viewed as essential not only to the ethical conduct of clinical trials, where patients are assigned to different therapeutic arms or to treatment vs. placebo, but also to determining the ethics of innovative interventions in a nonrandomized fashion. Maternal-fetal surgery for repair of fetal spina bifida is an example of innovative treatment that is experimental because its therapeutic effectiveness is not proved. If its effectiveness were proved, the treatment would appropriately be considered "standard of care," which as such should be offered to medically suitable candidates regardless of their ability to pay.

POSSIBILITIES FOR ETHICAL RESEARCH ON MATERNAL-FETAL SURGERY FOR REPAIR OF SPINA BIFIDA

Obviously, a complicating feature of the ethical requirement of equipoise in research on pregnant women is the presence of the fetus, whose interests, in terms of quality of life or survival, may be in conflict with the woman's interests. How can a particular researcher/clinician or group of researcher/clinicians be "equally poised" in their beliefs about the potential risks and benefits of treatment that carries no potential medical benefits but only risks for the woman? Even if equipoise were clearly established for the potential child, or if therapeutic benefit for the potential child were clearly demonstrated, medical equipoise for the woman herself would still be impossible.

The only way, I believe, by which equipoise can be established in these cases is by calculating risks and benefits in non-medical as well as medical terms.[14] Clearly, maternal-fetal surgery has implications for women, their fetuses, and their families well beyond those that would be considered physical or medical. If the medical risks could be counterbalanced by psychosocial gains, the ethical requirement of equipoise may be met. Psychosocial gains to women are likely because the women considering maternal-fetal surgery are usually the primary caregivers of children, and caring for children with disabilities often affects their psychosocial health in profound and potentially measurable ways.[15] If a woman's burden of caregiving were reduced by prevention, cure, or amelioration of her child's disability, that benefit might justify the harm she inevitably suffers by undergoing maternal-fetal surgery. It needs to be acknowledged, however, that women's burden of caregiving may be reduced in other ways than by surgery on them, e.g., by greater economic and social supports for care of the disabled.

To obtain data relevant to assessment of the psychosocial impact of maternal-fetal surgery, several areas of inquiry could be pursued. How, for example, do the burdens induced by long term caregiving and financial costs for women who undergo mater-nal-fetal surgery compare with those of women who elect standard therapy, and how do these data compare with the impact on women of decisions to terminate their preg-nancies in such circumstances? How do decisions that women make regarding mater-nal-fetal surgery impact upon their families? What are the burdens and benefits to them of gestational intervention, postnatal intervention, or termination? All of these questions deserve answers to ensure that equipoise is achievable in conducting research on maternal-fetal surgery.

From a policy standpoint, experimental interventions on pregnant women cannot be adequately assessed by focusing either on the fetus exclusively or on both the woman and fetus as separable or separate patients.[16] Instead, policy makers need to acknowledge that there is actually only one patient – the pregnant woman.[17] Just as any intervention in behalf of the fetus affects her, any intervention in behalf of the pregnant woman affects the fetus also. Operating on a woman to correct an anomaly in her fetus may be accurately described therefore as simply that: operating on *her*. As with decisions of other competent, informed adults regarding their treatment, her calculation of the risks and benefits necessarily includes assessment of the impact of her decision on those to whom she is related, and these, typically and appropriately, include partners and children already born. In light of the obstetric risks for future pregnancies, her calculation may also include future children.

Regarding the possibility of equipoise, the sole fact that a woman requests inno-vative treatment involving significant medical risks as well as potential psychosocial benefits to her does not suffice as grounds for equipoise on the part of the researchers themselves. On an account of the pregnant woman as the only research subject, the researchers only attain equipoise when they are genuinely uncertain regarding the *total* benefit/risk ratio of maternal-fetal surgery as compared with the alternatives of pregnancy termination and post-birth repair. At that point, a randomized clinical trial may be ethically pursued. Preliminarily, the questions I've posed above would have to be answered with data indicating that the potential medical and non-medical risks for the woman and the potential child are approximately balanced. Until and unless such data become available, the surgery should only be offered on an elective basis.[18]

SANCTITY OF LIFE VS. QUALITY OF LIFE CONSIDERATIONS

Early attempts to repair fetal anomalies in utero were mainly undertaken for conditions expected to be fatal for the newborn. For example, women whose fetuses were diagnosed with congenital diaphragmatic hernia, sacrococcygeal teratoma, and cystic adenomatoid pulmonary lesions were operated on after making the fetus accessible to surgeons through abdominal and uterine incisions.[19] In each case, it was hoped that, after the surgery, the pregnancy would continue to term and the newborn would survive. In other words, the calculation that underlay these attempts was the remote possibility of newborn survival through corrective surgery vs. certain death; or, in other words, small chance vs. no chance at life beyond birth: a sanctity of life rationale.

In attempts to treat non-lethal fetal conditions, the calculation that a small chance of survival justifies the intervention because the fetus cannot otherwise survive is not applicable. Regardless of whether such conditions are identified prenatally, most affected fetuses are delivered at term and treated as neonates in accordance with standard of care procedures. An irony unique to these prenatal procedures, as contrasted with prenatal surgeries for lethal conditions, is that they introduce mortality as well as morbidity risks that would not otherwise be present. Myelomeningocele or spina bifida is a prime example of a prenatally detectable, non life-threatening condition. Its incidence or severity can be reduced through folic acid intake before and during pregnancy.[20] According to bioethicists who have interviewed couples contemplating maternal-fetal surgery for spina bifida, the great majority of those who seek the surgery are opposed to abortion on moral grounds; they thus explicitly endorse a pro-life or sanctity of life ethic.[21] Despite this commitment, their decision to undergo treatment prenatally rather than after birth is based on quality of life considerations. Their willingness to impose an avoidable mortality risk on the fetus thus illustrates a greater aversion to life with disability than to loss of life.

Admittedly, "quality of life" is a problematic term. The elements that it entails differ for individuals and groups, making it impossible for one person to determine what constitutes quality of life for another. Public policies have at times warned against use of quality of life criteria in determining whether treatment of patients should be initiated or continued. The rationale for this concern is the subjectivity of the determination, allowing imposition of one person's or group's values on another. The same concern underlies the argument that use of quality of life criteria in treatment of people with disabilities disregards the fact that their interests may be different from those of their families or health professionals.[22] Despite these concerns, clinicians and the public in general often argue that quality of life criteria are important determinants of whether treatment should be provided, particularly when there is no way of accessing the patient's own wishes in that regard. For example, the American Medical Association specifically asserts that "quality of life, as defined by the patient's interests and values, is a factor to be considered in determining what is best for the individual."[23] (AMA Code E-2.17, 2.215) That quality of life criteria are relevant to physicians who provide maternal-fetal surgery is evident in their advocacy of mortality-risking surgery for nonlethal conditions. It is also evident in one of the exclusionary criteria for providing the surgery: prenatally detected nonlethal anomalies other than spina bifida.[24]

Quality of life considerations undoubtedly play an important part in public priorities and social attitudes about prenatal testing and termination of affected fetuses. Economic considerations play a part as well; these are relatable to quality of life factors because affluence can often buy a higher quality of life, and poverty generally leads to a reduced quality of life. Quality of life considerations are clearly relevant to the personal decisions of women and couples about prenatal testing and termination. Some have argued that use of these criteria illustrates the ableist view that persons already born with disabilities are not of equal value with those who are able.[25] If social supports for raising children with disabilities were adequate, I believe it is possible to avoid ableism while also defending the right to prenatal decision-making.[26] In our current social climate, however, those supports are lacking. Although individuals may reject the public priorities they learn from their social milieux, their values are surely influenced by them, often inadvertently but sometimes so significantly that the public priorities determine their personal priorities.

At times, the rationale that attaches to decisions about prenatal testing and termination alludes to the expectation of reduced quality of life in the affected newborn. The argument here is that postnatal life with a particular disability would be worse for the affected child than that he or she had never been born. Although a similar argument may be offered in defense of active euthanasia, it cannot be supported on grounds of the potential child's own wishes, as may be claimed for cases of active voluntary euthanasia. However, if the condition to be avoided is so severe that survival would in fact be worse than death for the affected child, the argument is comparable to one that supports active nonvoluntary euthanasia, i.e., killing an individual whose consent cannot be obtained on grounds of a duty of beneficence or nonmaleficence to him or her. Because suffering and pain tend to be intermittent or at least relievable with proper medication, there are few conditions for which this argument is persuasive – so long as adequate care is provided.

Quality of life criteria are even more problematic when they are applied to individuals whose disabilities apparently entail little pain or suffering for them. Down's syndrome, the principal condition for which prenatal testing and termination are undertaken, is a particularly challenging example in this regard. All children born with this condition are mentally retarded, but the degree of retardation varies greatly and is not predictable prenatally.[27] Although those affected are also more likely to have cardiac anomalies, leukemia, and other medical problems, some studies indicate that they tend to have pleasant dispositions, and seem at least as happy as their siblings.[28] However, if the quality of life referred to is that of the potential parents, whose caregiving responsibilities generally escalate when they have a child with disabilities, the claim that such considerations justify testing and termination is defensible. It may also be supported on grounds that responsibility to care for a disabled child compromises the quality of life of siblings who are not disabled, whose parents have fewer resources to care for them and who may themselves be expected to participate in the care of their affected sibling.

Another important difference between prenatal termination and active euthanasia based on quality of life considerations is that the individual who is euthanized is indisputably a person, both legally and morally, while fetal personhood is not legally established and remains morally controversial. Those who claim that the fetus has no moral status may support prenatal termination while condemning postnatal termination – for any reason; if they attribute some moral status to the fetus, the right to prenatal termination for any reason is more difficult to defend.

Despite its legality, abortion continues to be a politically charged and morally debated issue in the United States and elsewhere. Most abortion decisions, however, are not based on quality of life criteria applicable to the potential child; instead they are mainly defended on grounds of the woman's right to choose not to have a child, regardless of whether the child is able or disabled. Most abortions occur before prenatal testing is performed. The pregnant woman's decision may be influenced by considerations relevant to her own quality of life and that of others to whom she has caregiving responsibilities, but the pivotal consideration is her right to choose. In other words, the fetus has no moral status or "right to life" that could justify overriding a pregnant woman's decision to end its life. Most women who choose abortion thus reject a sanctity of life rationale; they do not necessarily base their decisions on quality of life considerations relevant to the potential child.

In contrast, women who seek maternal-fetal surgery for repair of spina bifida support a sanctity of life rationale by continuing their pregnancies. They oppose abortion because they believe the fetus has a right to life based on its moral status; regardless of whether their pregnancy was intended or not, their right to choose does not supersede that right to life. Moreover, like most pregnant women who reject the option of abortion, they are willing to do whatever they can to support the life of the developing fetus.

The problem with this pro-life or sanctity of life rationale is that it is inapplicable to maternal-fetal surgery after prenatal diagnosis of spina bifida. Attempting to repair the condition before birth not only fails to support the life of the developing fetus; it also introduces mortality and morbidity risks other than spina bifida. Even if (as is not the case) the major impairments associated with myelomeningocele were reduced in the newborn, deliberately introducing these risks by performing the surgery before birth cannot be justified on sanctity of life grounds. Consistency with a sanctity of life ethic after positive prenatal diagnosis would require continuation of the pregnancy and repair of the open lesion after birth.

Given this inconsistency, why is it that so many women who are morally opposed to abortion seek the surgery? And how do public priorities either merge or diverge with personal priorities with regard to this issue? In the remainder of this essay I examine these questions, which are unavoidably connected.

PERSONAL VS. PUBLIC PRIORITIES?

Consider the following possible reasons why women committed to a sanctity of life ethic choose a procedure that risks loss of life for their potential child

- ignorance of, or failure to acknowledge, the mortality risk of maternal-fetal surgery
- ignorance of, or failure to acknowledge, the morbidity risks of maternal-fetal surgery
- perception of an obligation to undergo any treatment that might promote the quality of life of their newborn
- encouragement of partners, other family members, and clinicians to undergo the procedure

- desire to avoid possible reduction in quality of life of themselves or other family members
- acceptance of, or acquiescence in, societal discrimination against people with disabilities.

Bioethicists who counsel women and couples considering maternal-fetal surgery for myelomeningocele report that they spend considerable time informing them of the mortality and morbidity risks of the procedure, as well as the fact that the surgery has never yet resulted in improved mobility or continence.[29] The couples persist, nonetheless, in their desire for intervention prior to birth. Most couples are married, and husband and wife are counseled together even though only the woman can actually undergo the surgery. In light of the discrepant impact on male and female partners, it has been recommended that they be counseled separately as well as together. In response to this suggestion, one member of the Vanderbilt team indicated that separate counseling would be strongly resisted by the couples themselves.[30]

The great majority of women who carry their pregnancies to term are willing and eager to undergo risks, inconveniences and even harms to themselves in order to promote the welfare of their potential child. Their decisions are typically guided by a genuine desire to improve their offspring's quality of life. Occasionally, however, social expectations and public priorities function coercively in this regard. Physical coercion of pregnant women is documented in reports of their detention in hospitals and in prisons for not conforming to behavior recommended for the sake of their fetuses. It is also evident in cases of coercive cesarean section.[31] Beyond overt cases of coercion, however, family pressures may compromise the autonomy of individual women to pursue their personal priorities with regard to fetal welfare. Women who are financially or emotionally dependent on their husbands, for example, may feel compelled to subject themselves to the risks of experimental treatment to which they would not agree if they were economically or psychologically independent. To the extent that family, cultural, or public expectations induce women to subject themselves to risks or harms for the sake of the fetus, those same influences may move them to undergo maternal-fetal surgery, despite its unproved benefits to the potential child.

Although mortality risks for the fetus escalate through maternal-fetal surgery, fetal demise due to the surgery has not been reported. The rise in the mortality risk is based on documented morbidity that could, if severe enough, lead to fetal loss or newborn death. Unlike treatment for otherwise lethal conditions, no lives are saved through this procedure. In fact, children born after maternal-fetal surgery for spina bifida may credibly claim that their parents deliberately endangered them by subjecting them to risks that they would die before birth. This claim is similar to the position supported by advocates of coercive treatment of pregnant women whose behavior, allegedly, places their potential children at risk. Of course, the parental motive in these cases seems laudatory: to reduce or avoid the disability with which the child would be born; in other words, to promote the child's quality of life. Given the risk of loss, however, deliberately undertaking it inevitably subordinates their sanctity of life commitment to quality of life considerations. In the context of a sanctity of (fetal) life commitment, it is comparable to parental desires to involve their children in research for which equipoise cannot be obtained. If the child's interests have priority over parental autonomy, this is neither legally nor morally justifiable.

Admittedly, people often take risks that they or others may die, when the risk is intended to promote their own or others' quality of life. When we drive rather than walk to a particular destination, for example, we expose ourselves to greater risk of accident and lose the health benefit of walking, while (presumably) advantaging our quality of life in other ways, such as less fatiguing or faster transit. And when adults undergo cosmetic surgery intended to make them look younger, they expose themselves to medical risks that are clearly avoidable. Maternal-fetal surgery for treatment of spina bifida differs importantly from these or similar examples of risks deliberately undertaken to improve quality of life. In the examples, real and desired benefits are legitimately judged proportionate to, or minimal in comparison with, the risks involved. The only benefits alleged to result from maternal-fetal surgery for spina bifida (reduced need for shunting and reduced hindbrain herniation) are not the principal goals of treatment (improved mobility and continence), and the mortality and morbidity risks to the potential child and to the pregnant woman hardly seem proportionate to, let alone minimal in comparison with, the benefits obtained.

The dilemma faced by women opposed to abortion whose fetuses have spina bifida is undoubtedly influenced by public priorities and attitudes about people with disabilities. Despite the illegality of discrimination and social efforts to ameliorate the disadvantages experienced by people with disabilities, discrimination against them has not only occurred throughout history but continues in our day and in our country. Sometimes the discrimination is deliberate, but indeliberate discrimination is probably more prevalent. Social structures are built by currently able individuals to accommodate their own needs and interests, and such individuals are often nearsighted with regard to the needs of people with disabilities.[32] Women whose fetuses are diagnosed with spina bifida are likely to recognize this myopia, and lament it because it will probably limit the opportunities afforded to their affected offspring and to themselves as primary caregivers of people with disabilities. By encouraging prenatal testing and termination of affected fetuses and by providing little support for those who are disabled, society in general gives priority to quality of life criteria over sanctity of life criteria, and women whose potential children are disabled cannot help but be affected by this prioritization, whether consciously or not.

If public priorities favor quality of life criteria, it may be argued that these criteria should have priority in personal decisions as well. In other words, when one faces a conflict between one's separate interests and those of the larger society, the latter should prevail. Obviously, this priority of the public over the private, which assumes a distinction between the two domains, has long been supported by political theorists and activists.[33] As Aristotle observed in the *Politics*, "The state is by nature clearly prior to the family and to the individual, since the whole is of necessity prior to the part." The evidence for this priority, he claimed, is that "the individual, when isolated, is not self-sufficient" but depends on the state.[34]

Although Aristotle is often cited as an advocate of "the public-private distinction," his assertion of the dependence of individuals on the state also provides grounds for critiquing those who interpret the distinction as a dichotomy. The "common good" that he champions has priority over some individual goods, but it is in fact a good for all individuals as such. Moreover, just as individuals depend on the state, the state depends on the individuals who compose it, and differences among those individuals contribute to its vitality and productivity. Inevitably, therefore, public priorities mix with personal priorities. For any individual, they overlap without being congruent; public priorities include only but not all personal priorities.

A similar and overlapping complementarity occurs, I think, between quality of life and sanctity of life criteria. All living individuals probably have some qualitative feature, at least to those who care for them, and quality of life cannot be maintained or developed without life itself being affirmed. In choosing maternal-fetal surgery for repair of spina bifida rather than waiting until birth so that risk of fetal demise through the intervention is avoided, women invoke quality of life criteria that they identify, rightly or wrongly, with their sanctity of life commitment. In choosing to terminate pregnancy after fetal diagnosis of spina bifida, women assess the quality of fetal life as so compromised that it does not warrant preservation through continuation of pregnancy. In both instances, the quality and sanctity of life of the woman herself is affirmed; in many cases, the quality and sanctity of others' lives may also be affirmed.

Acceptance of, or acquiescence in, societal discrimination against people with disabilities, illustrates personal priorities based on quality of life criteria. To the extent that such discrimination occurs, it fails to respect the sanctity of every individual life, whether able or disabled; in the case of prenatal testing and termination, it subordinates the value of fetal life to the value of women's autonomy. If and when decisions about abortion or treatment of fetuses are not made on grounds of society's overall unwillingness to value and care for people with disabilities, ableism will no longer prevail in our public and personal priorities.

In sum, maternal-fetal surgery for repair of myelomeningocele is an experimental treatment of pregnant women that can only be ethically justified by considering non-medical risks and benefits of the procedure and by avoiding impediments to autonomy in the decisions of women who undergo the treatment. Research on maternal-fetal surgery is needed to insure that it can be safely offered to all of those who may benefit from the procedure. Personal decisions to seek the procedure are undoubtedly influenced by public priorities, and sanctity of life considerations mingle with quality of life considerations in those who have already sought and obtained it. Critique of dichotomous interpretations of quality vs. sanctity of life as well as personal and public priorities seems crucial to honest and adequate ethical analysis of maternal-fetal surgery for treatment of spina bifida. Once different sides of the apparent dichotomies are recognized as overlapping, covert discriminatory attitudes and practices towards people with disabilities may be uncovered and addressed by individuals and by society at large.

NOTES

[1] Susan Dwyer, "Understanding the Problem of Abortion." in *The Problem of Abortion*, ed. by Susan Dwyer and Joel Feinberg. (Belmont, CA: Wadsworth Publishing Company, 1997): pp. 2-7.

[2] Because practitioners themselves tend to use the terms "spina bifida" and "myelomeningocele" interchangeably, I shall do so also. Cf. Anne Drapkin Lyerly, Elena A. Gates, Robert C. Cefalo, and Jeremy Sugarman, "Toward the Ethical Evaluation and Use of Maternal-Fetal Surgery." *Obstetrics and Gynecology* 98.4 (October 2001): 689-97, 692.

[3] George J. Annas, *Standard of Care: The Law of American Bioethics*. (New York, NY: Oxford University Press, 1993), p. 4.

[4] Anne Drapkin Lyerly, Robert C. Cefalo, Michael Socol, Linda Fogarty, and Jeremy Sugarman, "Attitudes of Maternal-Fetal Specialists concerning Maternal-Fetal Surgery." *American Journal of Obstetrics and Gynecology* 185.5 (November 2001): 1052-8, 1052.

[5] On a website (www.fetal-surgery.com) not affiliated with Vanderbilt University Medical Center, Joseph Bruner describes maternal-fetal surgery for myelomeningocele as "a new third option to either having an abortion or doing nothing."

[6] Joseph P. Bruner, William O. Richards, Noel B. Tulipan, and Timothy L. Arney, "Endoscopic Coverage of Fetal Myelomeningocele in Utero." *American Journal of Obstetrics and Gynecology* 180 (1999): 153-158, and Oluyinka O. Olutoyee, and N. Scott Adzick. "Fetal Surgery for Myelomeningocele." *Seminars in Perinatology* 23.6 (December 1999): 462-473.

[7] Hindbrain herniation is a malformation that may obstruct the flow of cerebrospinal fluid, causing hydro-

cephalus in 85% of individuals with myelomeningocele. Shunting is a procedure through which the accumulating fluid is diverted to absorbing or excretory systems. Cf. Sutton, Adzick, Bilaniuk et al., and Bruner, Tulipan, Paschall et al.

[8] Joseph P. Bruner, Noel Tulipan, Ray L. Paschall, Frank H. Boehm, William F. Walsh, Sandra R. Silva, Marta Hernanz-Schulman, Lisa H. Lowe, and George W. Reed, "Fetal Surgery for Myelomeningocele and the Incidence of Shunt-dependent Hydrocephalus." *Journal of the American Medical Association* 282.19 (November 17, 1999): 1819-1825.

[9] www.fetalsurgeons.com. The other two Centers that provide maternal-fetal surgery for spina bifida are at the University of California, San Francisco (www.fetalsurgery.ucsf.edu), the longest-standing program, and Children's Hospital of Philadelphia (www.fetalsurgery.chop.edu), whose practitioners were trained at UCSF. All three programs draw patients from their websites. The latter two are led by pediatric surgeons, whereas the former is led by an obstetrician and a pediatric neurosurgeon.

[10] The conference at Vanderbilt was held in Nashville, TN, on March 11, 2000; the one at the National Institutes of Health was held in Bethesda, MD, on July 16, 2000. The author participated in both conferences.

[11] The acronym for this multi-center study is MOMS (cf. "Management of Myelomeningocele"); a description of it is available at http://www.spinabifidamoms.com/overview.html.

[12] During the discussion of the topic at the two conferences mentioned above, Dr. Bruner expressed the opinion cited in the text. A feature of the multi-center trial that will facilitate recruitment is that potential patients do not pay to participate in the research. Prior to the trial, patients typically paid for the procedure out of pocket because, as an experimental procedure, it was not covered by their health insurance. In contrast, surgical repair of spina bifida after birth is covered by most insurance polices because it is viewed as standard of care.

[13] Nicolas Johnson, Richard J. Lilford, and Wayne Brazier, "At What Level of Collective Equipoise Does a Clinical Trial Become Ethical?" *Journal of Medical Ethics* 17 (1991): 30-34.

[14] Anne Drapkin Lyerly, and Mary B. Mahowald, "Maternal-Fetal Surgery: The Fallacy of Abstraction and the Problem of Equipoise." *Health Care Analysis* 9.2 (2001):151-65.

[15] Mary B. Mahowald, Dana Levinson, Christine Cassel, Amy Lemke, Carole Ober, James Bowman, Michelle Le Beau, Amy Ravin and Melissa Times, "The New Genetics and Women." *The Milbank Quarterly* 74. 2 (1996): 239-83, 268-270.

[16] Mary B. Mahowald, "As If There Were Fetuses without Women," *Reproduction, Ethics and the Law: Feminist Perspectives*, ed. by Joan C. Callahan. (Bloomington, IN: Indiana University Press, 1995), pp. 199-218, 199.

[17] Lyerly and Mahowald, p. 163.

[18] The same rationale applies to cosmetic surgery that entails medical risk not associated with any expected medical benefit. Research is necessary to establish the relative safety of the procedure despite its inevitable health risks. In both cases, once safety concerns have been resolved satisfactorily, practitioners are permitted but not obliged to perform the surgery. However, an ethically relevant difference between these cases is that the fetus cannot consent to the surgery.

[19] Lyerly, Gates, et al., p. 690.

[20] To the credit of organizers of the Vanderbilt conference, the first day of the two-day meeting was devoted to this preventive measure.

[21] Mark J. Bliton, and Richard M. Zaner, "Over the Cutting Edge: How Ethics Consultation Illuminates the Moral Complexity of Open-uterine Fetal Repair of Spina Bifida and Patient's Decision Making." *Journal of Clinical Ethics* 12.4 (2001): 357.

[22] *Federal Register* 49. "Nondiscrimination on the Basis of Handicaps; Procedures and Guidelines Relating to Health Care for Handicapped Infants." (January 12, 1984): 1622-1654.

[23] In medical decisions for seriously ill infants, the AMA Code says "the anticipated quality of life for the newborn with and without treatment" should be weighed "from the child's perspective." (section E-2.215)

[24] At the Vanderbilt conference, Dr. Bruner was asked whether the surgery would be offered to a woman whose fetus had Down syndrome. When he did not respond affirmatively, it was suggested that a refusal could be viewed as discrimination against the disabled, i.e., ableism.

[25] Parens and Asch call this "the expressivist argument," whose central claim is that prenatal tests to select against disabling traits "express a hurtful attitude about and send a hurtful message to people who live with those same traits." Erik Parens and Adrienne Asch, "The Disability Rights Critique of Prenatal Genetic Testing." *Prenatal Testing and Disability Rights*, ed. by Parens and Asch. (Washington, D.C.: Georgetown University Press, 2000), p. 13.

[26] Mary B. Mahowald, *Genes, Women, Equality*. (New York, NY: Oxford University Press, 2000), pp. 149-50. Compatibility between advocacy for persons with genetic disabilities and advocacy for women's right to prenatal testing and termination requires that the mere fact of the disability is irrelevant to the choice. Other reasons, such as lack of resources to care for a disabled child and/or the probability that the child's life will preponderantly be one of suffering or pain, may justify testing and termination without constituting a discriminatory attitude on the part of the woman.

[27] Ricki Lewis, *Human Genetics: Concepts and Applications*, 2nd edition. (Dubuque, IO: Wm. C. Brown Publishers, 1977), p. 210.

[28] Kenneth L. Jones, *Smith's Recognizable Patterns of Human Malformation*. 4th ed. (Philadelphia, PA: W. B. Saunders, 1988), p. 11. P. Gunn and P. Berry. "The Temperament of Down's Syndrome Toddlers and Their Siblings." *Journal of Child Psychology and Psychiatry* 26.6 (1985): 973-9, 973. Brian E. Vaughn, Josefina Contreras, and Ronald R. Seifer, "Short-term Longitudinal Study of Maternal Ratings of Temperament in Samples of Children with Down Syndrome and Children Who Are Developing Normally." *American Journal of Mental Retardation* 98.5 (1994): 607-18, 607.

[29] Bliton and Zaner, p. 357.

[30] The person who made this comment participated in the discussion after one of the talks at the Vanderbilt conference. She did not give her name, but identified herself as a member of the Vanderbilt team involved in counseling potential patients.

[31] Veronika E. Kolder, Janet Gallagher and Michael T. Parsons, "Court-Ordered Obstetrical Interventions." *New England Journal of Medicine* 316 (1987): 1192-96.

[32] "Myopia" is my favorite metaphor for the flaw that characterizes all human beings: each of us can only see the world from where we stand in it, and that means that we cannot see all that there is to see or see it clearly. The corrective lens needed to reduce our inevitable nearsightedness is to listen to what

those who stand in different places see from their perspectives. By incorporating perspectives other than our own, particularly from those whose standpoints tend to be ignored because they are non-dominant members of society, we can achieve a clearer, less inadequate account of reality. Mary B. Mahowald, "On Treatment of Myopia: Feminist Standpoint Theory and Bioethics." *Feminism and Bioethics: Beyond Reproduction*, ed. by Susan M. Wolf. (New York, NY: Oxford University Press, 1996), pp. 95-115, 96-98, cf. Anita Silvers, David Wasserman, and Mary B. Mahowald. *Disability, Difference,Discrimination: Perspectives on Justice in Bioethics and Public Policy.* (Lanham, MD: Rowman & Littlefield Publishers, Inc., 1998), p. 4.

[33] Okin, however, has criticized the distinction because it tends to provide "cover" for inequities that occur in non-public domains such as the family. Susan Moller Okin, Justice, Gender, and the Family (New York, NY: Basic Books, Inc., 1989), pp. 128-33. And Young offers a more radical critique, arguing that the distinction "masks the ways in which the particular perspectives of dominant groups claim universality, and helps justify hierarchical decision making structures." Iris Marion Young, Justice and the Politics of Difference (Princeton, NJ: Princeton University Pres, 1990), 97.

[34] W. D. Ross trans., *The Pocket Aristotle.* (New York, NY: Washington Square Press, 1958), p. 281.

MICHAEL BOYLAN

GUN CONTROL AND PUBLIC HEALTH

ABSTRACT: This essay will explore the way we should think about the ethical and public health implications of gun control in the United States of America today. The generating pedagogy will be: 1. An explication of Worldview perspectives—both personal and community as per the author's recently published writings. 2. A discussion of the worldviews of both sides of the gun control debate. 3. A critical appraisal of the positions of each side. 4. Some suggestions about a future that is without ordinary citizen ownership of guns. This future would deny ordinary citizens their right to bear arms because this right is superceded by a more fundamental right that is connected to the public health of the United States.

KEY WORDS: Gun Control, Public Health, worldview

Each day the emergency room in countless hospitals across the country is beset by young men and women who are admitted with severe gunshot wounds that require surgical care. Why is it that some of our major cities have as many as one shooting (on average) each day?[1] How can we make sense of this steady stream of carnage? What should be done about it?

The phenomena of gunshot victims, the psychological pain and agony of family members and the general health and well being of members of communities in which gun violence is prevalent, make gun violence a public health problem that cannot be ignored.

This essay seeks to explore the phenomenon of gun violence in the United States and then explore the worldviews of gun ownership advocates as well as gun control advocates with an aim to discover what social action ought to be taken.[2]

There are many versions of gun control including gun registration, gun education classes, and the like. To draw the lines more sharply between those who advocate gun ownership and those who do not, this essay will understand gun control to mean that ordinary citizens, *ceteris paribus*, will not be allowed to own a gun. The only exception to this restriction would be when one can demonstrate a fundamental need for a gun (more on this later).

119

M. Boylan (ed.), Public Health Policy and Ethics, 119-134.
© 2004 Kluwer Academic Publishers. Printed in the Netherlands.

THE PHENOMENA

Each year there are over 28,000 fatal gunshot victims in the United States. The causes for this phenomenon are very complex and defy full exposition in this essay. One set of causes is social/political in origin. These factors revolve around the agents themselves and why they decide to carry guns. These causes will be addressed later in this essay.

Another important concept focuses upon the weapon itself. What makes one weapon preferable to another and what does it mean to carry one sort of weapon as opposed to another? One way to get a handle upon this important concept is the notion of a weapon's *damage coefficient*. For the purposes of this essay, a damage coefficient may be assigned to a weapon in proportion to the following factors: (a) the normal extent of damage that can be expected to be caused by a weapon in its most efficient application (on a scale of 0-10); (b) the probability that the most efficient application will take place (on a scale of 0-1). To arrive at a weapon's damage coefficient, we multiply (a) times (b). For example, let there be two weapons, a and b. a normally could kill in an efficient application (10) but it rarely would happen (.1). The damage coefficient for a would be 1.

In contrast, b could normally only bloody your nose (2) and it almost always happens (1). The damage coefficient for b is 2. Thus, on this account, the damage coefficient of b is greater than a.

The damage coefficient is an important concept in evaluating weapons because it allows us to balance the normal potential for the weapon against that potential's probability of occurring. It is the contention of this essay that a weapon that possesses a higher damage coefficient is more *dangerous* in practice. The damage coefficient, of course, weighs practice over potentiality. This is especially useful when one thinks in group dynamics, but it does not reveal the entire story. A lethal weapon (such as an Eighteenth Century dueling pistol) may have a very low effectiveness and thus a low damage coefficient, but it still may cause fear and threaten another for its potentiality—what it *might* do. This potentiality creates a palpable effect that should not be ignored. As much as possible, this essay will try to keep both sorts of concerns in mind as we evaluate weapons.

When we consider modern weapons, it is clear that guns are a very unique sort of weapon. To understand their uniqueness one must imagine the world before there were guns. In this world the most common weapon in domestic situations was probably the knife, fist, or physical contact with a blunt object—that might include extensions of the fist such as an implement of pottery. Thus, if Jake got mad at John and wanted to do violence to him because he suddenly became angry, the most common weapon (unless he were in the kitchen) would be his fist or some common implement that was at hand such as a plate or vase. Now these latter implements can be lethal in their own way—especially if the blow is delivered in an especially sensitive spot on the head, spine or some other area of vulnerability. But these weapons have a much lower damage coefficient than does a modern pistol or rifle. This is because when one uses a firearm as his weapon, the probability of severe bodily harm or death approaches 1 while earlier weapons (including knives) had a much lower probability. This lower probability is largely due to the weapon having a lower inherent danger. In order to kill with one's fists, a flowerpot, or even a knife, some skill is involved on the attacker's part or bad luck on the victim's. However, the victim of the attack also has more opportunity to avoid bodily harm in the case of an attack by fists, flowerpot, or even knife (than by an attack by a gun). The victim's odds in the conventional attack are even better if she is a skillful street fighter.

But even the most successful street fighter falls to the pavement when shot with a gun. The accuracy and ease of operation make the gun an "equalizer." Small and weak people can kill as effectively as savvy street fighters. This fact makes the gun the weapon of choice for all those seeking an offensive or defensive weapon. The amount of damage that normal bullets create (not to mention the increased amount of damage that split or exploding bullets wreak) makes the modern gun clearly the most efficient device for tearing, shattering, and destroying human tissue, bone, and vital organs. This conjecture is borne out by the statistics. In 1999 the United States Federal Bureau of Investigation listed 12,658 total murder victims. Of these 8,259 were killed by firearms. The next most lethal weapon was the knife, 1,667 victims. The third most lethal weapon was some sort of blunt instrument, 736 victims.[3] Therefore, for the purposes of this essay it will be assumed that guns (including rifles) have the highest damage coefficient of any common individual weapon.[4]

PERSONAL AND COMMUNITY WORLDVIEW.

How should people and communities confront the phenomena of modern guns and rifles? One way to approach this question is through the concept of worldview. An individual's worldview is an examined depiction of facts and values that the agent holds to be true.

The Personal Worldview Imperative is a normative command that everyone must undergo regular self examination of his worldview according to the highest standards, "All people must develop a single comprehensive and internally coherent worldview that is good and that we strive to act out in our daily lives."[5] The three principal conditions that are set upon a worldview are that it be comprehensive, coherent, and good. From the outset these conditions infuse an ethical viewpoint into everyone's daily life.

There are two conditions that attach themselves to the exposition of a personal worldview: sincerity and authenticity. The Personal Worldview Imperative dictates a self-examination by a sincere, authentic agent. This presupposes that agents possess the capacity of reflection and choice. If the agent reflects according to her highest abilities and is motivated to seek the truth, then the agent can be termed sincere.

If the agent strives according to the structures in the Personal Worldview Imperative (including all of her values—not merely ethical values) and is willing regularly to review her worldview according to the dialectical process advocated by the imperative[6], then the agent can said to be authentic. Thus, sincerity is about intent while authenticity is about intent within a proper structure. Neither sincerity nor authenticity is sufficient to produce a worldview that is coherent, complete, and good, but no one can will demand more of herself than this.

Why be sincere and authentic agents? The reason that it is good to be a sincere and authentic agent is that only sincere and authentic individuals fully actualize what it means to be an agent in the world. This is because only the sincere and authentic agent is truly autonomous, a self-law-maker.[7] Only an autonomous agent can said to be fully responsible (both positively and negatively) for what he does. Our human nature dictates that before all else, we desire to act. Ergo, our human nature dictates that we strive to be autonomous. If the only way we can be fully autonomous is to be sincere and authentic agents, then on the pain of violating our human nature we should strive to be sincere and authentic agents.

The second imperative concerns our roles as agents living in a society. This imperative is called the *Shared Community Worldview Imperative*, "Each agent must strive to create a common body of knowledge that supports the creation of a shared community worldview (that is complete, coherent, and good) through which social institutions and their resulting policies might flourish within the constraints of the essential core commonly held values (ethics, aesthetics, and religion)." There are several key elements to this imperative. First, there is the exhortation to create a common body of knowledge.[8] This is an essential element in order for positive group discussion to proceed. Second, there is a dialectical process of discussion among members of a single community and between members of various single communities that are united in another larger heterogeneous community. This discussion should seek to form an understanding about the mission of the community within the context of the common body of knowledge and the commonly held core values held by members of the community. These values will include ethical maxims, aesthetic values, and religious values. Of course, there will be disagreements, but a process is enjoined that will create a shared worldview that is complete, coherent, and good.

Third, is that the result of this dialectical creation of a shared community worldview is to employ it in the creation (or revision of) social institutions that are responsible for setting policy within the community/social unit. It should be clear that this tenet seems highly inclined toward democracy. It is. However, it is not restricted to this. Even in totalitarian states the influence of the shared community worldview is significant. One can, for example, point to the great differences among communist states in Eastern Europe, the Soviet Union, China, North Korea, and Cuba during the 1960' s-80. All were Communist. Yet there were great differences in the way the totalitarian regimes operated in each instance. This is because, even without the vote, the shared community worldview casts a strong influence upon the operation of society's institutions and their resultant policies.

Finally, it should noted that the actions of those institutions must always be framed within the core values of the people who make up the society. Whenever the society veers too far away in its implementation of the social worldview from the personal worldviews of the members of the society, then a realignment must occur. In responsive democracies this takes the form of changing the legislature in the next election. In totalitarian regimes, change will also occur, but generally by coup d'état or armed revolution.

These two principles work together to form the basis of individual claims to the goods necessary for agency. It takes the personal perspective (The Personal Worldview Imperative) to set out what counts as a legitimate rights claim, and it takes the community perspective (The Shared Community Worldview Imperative) to allocate those goods according to who has the strongest claims. But what makes one claim for a good stronger than another? To answer this question we need to add one more theoretical structure. This is the Table of Embeddedness (see "The Moral Basis of Public Health," p. xxiii). One good is said to be more embedded or fundamental according to its relation to committing action at all. The table begins with the most fundamental and moves to the less fundamental all the way to the superfluous.

The point of the Table of Embeddedness is to argue that the fundamental conditions that permit action are most primary to all agents. The goods of agency differ in their primacy to action. Those goods most primary are more important to action than those that are less primary.[9] It should also be noted that since level-one basic goods are justified because of their biological proximity to action and since medicine and public health are about maintaining a reasonable level of biological well being to individuals, then the maintenance of level-one basic goods becomes not only a moral and political imperative, but also an imperative of public health. Gun violence denies people of the basic goods of agency. Since the basic goods of agency are foundationally justified via an appeal to the biological foundations of action, and since all foundational biological needs are the province of medicine and public health, it follows that gun violence is a medical and public health issue.

THE WORLDVIEWS OF THE GUN OWNERS VS. THE GUN CONTROLLERS

With these distinctions in mind, let us explore the worldviews that have been created by both the gun owners and those advocating gun control (or abolition). First we will set out the worldviews as they exist and then we will critically evaluate these worldviews.

The Worldview Position of Gun Owners.
The gun is a tool. Like any other tool it can do good or evil. It all depends upon the person who owns the gun whether the tool will be properly used or not. Like any tool, a gun has certain primary functions. These include: hunting, protecting cattle, target shooting, and self-protection. Let's look at these in order.

Most hunters in the United States do so as a form of recreation. There are only a very very small number of people who hunt, fish, and trap in order to stay alive. For these individuals, a gun is a much more efficient weapon for killing animals than say a bow and arrow are.

For the recreational hunter, owning a gun allows them to pursue goods that are pleasant. These individuals do not need hunting in order to live or in order to pursue a plan of life that is essential to their actualization as people. Instead, the gun provides a means for a pleasant form of relaxation and recreation.

Since the beginning of time those who tend to cattle have had to protect the cattle from wolves and other predators. Though early shepherds used stones and slings as their weapon, the modern counterparts use guns. In some ways stones and slings are better. They do not kill the wolf (or other predator), but merely stun them and cause them to leave the cattle alone. In the terminology of this essay, stones and slings have a low damage coefficient.

However, in our increasingly ecologically unbalanced world, it may be the case that a weapon with a higher damage coefficient may be in order. This is due to the fact that the checks and balances of evolutionary nature are often skewed. Because of this imbalance, it may be impossible to keep these predators at bay by hurling rocks at them. Something else is needed: a weapon that can stop a greater number of animals might be necessary not only to save the livestock, but also to save the life of the farmer.

Target shooting is a recreation that many people consider to be very pleasant. To go to a club and test one's aim at various distances or to shoot skeet on a course are sports that are enjoyed by many Americans.

The last position that gun advocates support is self-protection.[10] The world is a scary place. There are many bad people out there who *do* own guns. Since the bad people do not obey laws anyway (otherwise they wouldn't be bad people), they will always own guns. The only way to avoid being an automatic victim is to own a gun that you either carry about with you as a concealed weapon or to have one stashed at home in a special place (such as your bedside table) so that you might thwart a home intruder who might rape and kill your family members.

Because a gun is a tool that can be used for good or for ill, it is the position of the gun owners that they should be allowed to own and carry guns.[11]

The Worldview Position of Gun Control Advocates.

Gun control advocates generally focus their worldview attitudes on the very high damage coefficient that guns possess as a weapon. This means that guns are uniquely dangerous. As such, they should be regulated or prohibited. For example, these individuals will contend that if Ms. X gets into an argument with her boyfriend Mr. Y and is sufficiently agitated, then Ms. X may decide to use a weapon against Mr. Y in order to sustain her strong opinion. If X hits Y with a frying pan, then Y will have a headache or possibly a concussion. However, if X uses a gun, she has the strongest weapon possible. But because guns have such a high damage coefficient, if X uses a gun she is more likely to kill or permanently injure Y, than if she had chosen any other weapon.

A very high percentage of murders (47.7%) are committed by people who are in the same family or who know each other.[12] Often, such crimes are committed in the passion of the moment. One party gets mad and searches for a weapon to vent her anger. If a gun is available, then because of the damage coefficient, the resultant harm will probably be far greater than if only a weapon with a lower damage coefficient were available.

Those who do not have guns are often fearful of those who do (lawfully and otherwise). For example, at my youngest child's school it was recently recommended that before you approve a play date for your child at another family's house, that you ask whether there are guns in the house and what precautions are in place to protect your child from a possible firearms accident.

From the point of view of those advocating gun control the entire tenor of the social environment changes when there are guns. In many ways, those without guns are almost as frightened by those who possess guns legally as those who don't. This group of people feels that the social environment will become safer all around if guns were removed from society.

A Critical Evaluation the Respective Worldviews.
1. Gun Owners. The gun owners were divided into four groups: hunting, protecting cattle, target shooting, and self-protection. If the subsistence woodsman needs to carry a gun in order to live, then he would see the need for owning a gun as some form of Basic Good (i.e., as a means of providing food, a level-one Basic Good).

The other form of hunter is one who does so because she finds it relaxing to go out into the woods and get back to nature. The recreational hunter takes up the sport because it is pleasant for her to do so. Thus, for the recreational hunter, the good of having a gun is for the purpose of pursuing a level-three Secondary good (a luxury good of low embeddedness).

Those who raise sheep and cattle for a living must protect their animals. As a result of human-generated pollution, the natural biome has been skewed. Predators and prey are artificially and differentially protected. Because of this, it may be the case that ecosystems are so out of balance that an unusually large pack of predatory animals might descend upon a shepherd. This might make the possession of a gun more than a tool to protect his chosen livelihood (a level-one Secondary Good), but even his life (a level-one Basic Good).

Target shooting is a pleasant recreation. As such, target shooting is a level-three Secondary Good (luxury—low embeddedness).

The last and most difficult group concerns those who wish to possess guns for safety. "Protection from unwarranted bodily harm" is a level-one Basic Good. If owning a gun allowed one to be protected in this way, then the claims right to gun ownership would be very strong. On the face of it these relations might be categorized as follows:

	B-1	B-2	S-1	S-2	S-3
1a Subsistence Woodsmen	x				
1b Recreational Hunters					x
2 Cattle Ranchers			x		
3 Target Practice					x
4 Safety		x??			

Table One: The Relative Embeddedness of Rights Claims by Gun Owners
(B-1, B-2 refer to basic goods levels 1 & 2; S-1-3 refer to secondary goods levels 1-3)

Each attribution in Table One seems to be solid except for the fourth category of safety. This is probably the largest group of gun owners so that it is important to be clear on this motivation for gun ownership. This essay has identified two sorts of people who wish to carry guns for protection: those who wish to carry a concealed weapon for the sake of personal protection (in what they feel to be unsafe social environments) and those who keep a gun at home so that they might protect themselves in the event that someone breaks into their house and might harm the family.

In the first case, one must assess the different mental attitude that one might have if he were to carry a concealed weapon as opposed to *not* carrying a concealed weapon. In my own experience I have never owned nor carried a weapon (even though I have lived in some of the highest murder-rate areas in the United States). Thus, I feel that I might be able to give a personal testimony for this group. I also know people I respect who (in similar circumstances) did buy a gun. Thus, I feel that I can represent this position (though from the stance of one who made a contrary choice).

In the first case, why would someone want to carry a concealed weapon on a daily basis? This person's worldview depicts reality as very hostile. There are people in the world, who in their pursuit of crime, would not think twice about killing you. Who should be the victim in such a confrontation: the criminal or you? Most would say the criminal. But how safe are you when you pack a concealed weapon? Imagine this (typical?) situation. Mugger X approaches good citizen Y. Because X wishes to be successful in his robbery attempt, he has his weapon already out so that at some moment of X's choosing he places the gun barrel against Y's head and declaims, "Your money or your life." Does the fact that Y has a concealed weapon in his inner coat pocket bring him safety? Let's continue with the scenario. If Y is reckless, he may reach under his coat and *pretend* to deliver his wallet when he is really taking out his gun. Y then reels and shoots X dead. Y feels that he has just thwarted a criminal.

But is the ending of that scene very accurate? If X is holding a gun to Y's head and Y makes the sort of sudden movements described, then isn't it just as (even more?) probable that X will now shoot Y before he can attempt the sort of heroics that are often depicted in the movies? If Y ends up dead because he couldn't perform the unrealistic feats depicted in the cinema, then is Y safer for having packed a concealed weapon? Certainly not.

I believe that it is generally the case that unless one is an individual who encounters and reacts to dangerous situations involving firearms on a very regular basis (such as police or army personnel do), that the possession of a concealed weapon will not enhance personal safety.[13] The individual may *feel* empowered, but that may have something to do with a fixation upon the gun and an appreciation of its intrinsic damage coefficient. If this is correct, then the possession of a concealed weapon may actually *impair* personal safety because it creates a false sense of security.

The second sort of case involves a person at home who possesses a gun in order to protect her household. This individual may keep a gun in her bedside stand so that if an intruder came into her house, she might be able to pull out the gun and defend herself and her family. In reality, if the person in question had a family, then she would not keep a loaded gun in her bed stand. To keep a loaded gun in one's bed stand when she has children is to invite tragedy.

Thus, this woman (being a responsible person) keeps an unloaded gun in her bed stand and has the ammunition locked away in her jewelry chest. Now let us suppose that a burglar breaks into her house and moves directly toward the bedroom. The time that it will take for the burglar to get to the bedroom is 60 seconds. The woman awakes, clears her head and reaches for her gun (30 seconds). She rushes to her top dresser drawer to get her jewelry chest key (15 seconds). She rummages about for her key (15 seconds). She opens up her jewelry chest and takes out three bullets (5 seconds). Then she loads her gun (15 seconds). Now she is ready for the home invader—20 seconds too late!

Obviously, this scenario is weighted against the possibility of success. But as a light sleeper myself, and having encountered a real home invasion on one occasion and many more "false alarms" I can attest that awakening from sleep is not conducive to quick and efficient behavior. The general point is that if one is responsible (i.e., keeping firearms unloaded with ammunition locked-up), then it is very questionable whether having a gun at home really enhances security.[14] Perhaps one might be more secure if she invested in a monitored burglar alarm system that would automatically call the police (who in most metropolitan areas will arrive on the scene within five minutes).

My conclusion is that though protection from unwarranted bodily harm is a basic good of agency, it remains to be proved whether owning a firearm for personal safety as either a concealed weapon or in one's home really performs as promised. Given that there are many safety issues at stake, it would seem that the risk does not justify the reward.[15]

2. Gun Control Advocates. Gun Control Advocates run the risk of being accused of exaggerating the risk that personal ownership of guns really poses in our society. The detractors to gun control cite scenarios that if the majority of people in the United States (or any other country) were armed, then the level of violence would decrease. This is because normal people on the street would realize that if they were offensive

or threatening in their behavior there might be a lethal consequence. This is akin to the M.A.D. (mutual assured destruction) theory that was employed by the United States and the former Soviet Union during the Cold War. In that case being armed actually seemed to bring about peace. On this model, advocates of universal gun ownership would tout the advantages of bringing justice down to the local level.

This theory also fits in with anti-central government sentiment and returning power to the local level and people, respectively. This is because arming the population would (by these proponents) empower the people. If a bad guy wants to try something with me, then he will have to deal with my bodyguards, Mr. Smith & Mr. Wesson.

However, I reject this option. I do not believe that we lived in a peaceful tranquillity during the Cold War. The only thing that stopped nuclear holocaust during that period, in my opinion, was the fact that *nations* and not isolated individuals were making the decisions. Nations generally make important decisions through a group consultation framework. This tends to reduce greatly the chance that some isolated individual might act precipitously. On the other hand, with a society of armed individuals, I believe that the potentiality for erratic and irrational action is far greater than it would be in a society aspiring to comprehensive gun control. On the individual model there are many instances of the gun being used to vent anger and despair. This marks the significant difference between the group and the individual model. This is why M.A.D. could have worked at the group level to *prevent* violence, but be an invitation to *increased* violence at the individual level.

My conjecture is partially based upon the widespread gun use in the inner cities of the United States. 'The inner cities' is a euphemism for places of poverty. In the United States in the beginning of the 21st Century, the contrast is stark between people who live in those geographical islands of few secondary goods and those in the mainland of affluent society. Human feelings of jealously and justice can incite rage among the dispossessed.[16] But it is a rage that is generally turned inward. It does not lash out at those who *have* or at those who *control* and *profit* from the poor. No, for the most part the rage is generated against compatriots who also live in poverty. The fact that poverty is also statistically correlated with certain racial and ethnic groups (such as African-Americans and Hispanic Americans) creates another factor in the equation: prejudicial exploitation.

Since poverty and crime often go hand in hand,[17] and since crime and firearm use certainly correlate together, then those in poverty will experience more gun violence than those living elsewhere. This brings about an unfortunate double consequence of living poor in the inner city, viz., not only is one deprived of the secondary goods that are generally owned by most people in the society, but one is environmentally deprived of a basic good of agency, i.e., protection from unwarranted bodily harm.[18] When these effects fall disproportionately upon one racial/ethnic group (such as African Americans and Hispanic Americans), then it would seem as if that group or groups were being singled out for unjustified exploitation. If this interpretation is correct, then the plight of those in the inner cities deserves a concerted societal response.

But what does this have to do with gun control? Are we making those in the inner city safer or more vulnerable by asking law-abiding citizens to turn in their guns?

DIRECTION FOR THE FUTURE

As argued above, I believe that law-abiding citizens are *not* safer when they own guns. I recognize that under the definition of gun control that began this essay (i.e., gun abolition), that weapons will be differentially in the hands of criminals who would have an advantage over the law abiding citizen in a robbery attempt. But this happens now. Since the overwhelming number law-abiding citizens who own guns don't regularly use their guns, they aren't practiced enough to be able to successfully confront a gun-wielding thug. The difference is that under gun control the ordinary citizen won't artificially rely upon an incorrect apprehension of his or her personal safety. They will seek other forms of risk control. In addition, the possibility for collateral accidental shootings is eliminated among law-abiding citizens.[19]

On the other hand, gun advocates might contend that if the issue of safety is unproven, why not side with those wishing to own guns—even if their wish (for safety) is not really borne out in facts. What does it hurt if individuals are allowed to exercise their autonomy?

This is an argument that on one level is about the burden of proof. Should we allow guns unless it is demonstrated conclusively that they do not improve personal safety or should we *not* allow guns unless it is demonstrated conclusively that they *do* improve personal safety (and have no further drawbacks). Imagine that there were no guns in the United States (or any other society for that matter). Now pretend that businessmen are walking forward asking for permission to allow guns to sold to private citizens. Does anyone really believe that it would pass muster? When children's toys are scrutinized because the toy gun on an action figure might be swallowed and cause choking, how does anyone really expect that real firearms would (in the modern age) ever be approved?

Perhaps this is unfair. The same argument could be made about tobacco (viz., that it would never be approved). But the fact is that the genie is out of the bottle. However, the thought experiment is not in vain. It indicates where the burden of proof ideally should lie (with the gun advocates).

Another sort of test is the *novus actus interveniens.*[20] Under this test, anything that alters the natural order must be justified because it is causally responsible for any and all ensuing consequences. Since guns are not a part of the natural order, they must be justified as interventions in that order. To do so means that the burden of proof falls upon the gun advocates.

If I am correct that the burden of proof falls upon the gun advocates, and if they can (at best) show that there *may* be some incremental (actual and not merely perceived) increase in personal safety through carrying a gun, then they may have an argument for the moral right to carry a gun. But this is only true if the incremental benefits outweigh the other collateral costs (i.e., the number of homicides among family members of otherwise law abiding citizens—which we remember is roughly half of all homicides). I do not think that gun advocates can do this on the issue of personal safety (as per my earlier arguments).

There may be scenarios in which certain occupational groups (such as cattle ranchers or those living in the wilderness) need to be able lawfully to possess a firearm. The society should make such accommodations. However, in the largest group of all—those owning guns for the sake of protection, I suggest that the perception is illusory. We are not safer when we possess guns. It is a cruel illusion. It is especially cruel among those most vulnerable in our society, those living in poverty. Because this lot falls disproportionally upon African Americans and Hispanic Americans, this cruel illusion has the result of increasing the burden of societal oppression upon these peoples.[21] This is graphically illustrated by the fact that in the most recent statistics available the rate for death due to firearms among African American Men is 40.3 per 100,000 as opposed to 17.4 per 100,000 among white Americans.[22] This makes African American men at 2.3 times the risk of death by firearms than their white counterparts. Guns hurt the poor the most. Therefore, they will be greatly benefited by a policy of gun control.

Thus, though individual applications of the personal worldview imperative might claim the right to own a gun due to the Basic of Good of protection from unwarranted bodily harm, I believe that factually this claim is mistaken and that safety is not materially enhanced. This conjecture is conjoined with the considerable societal harm that ensues from the availability of and widespread possession of guns among law-abiding citizens. If guns were only available to those who could document that a basic good of agency were at stake, then the dictates of the Shared Community

Worldview Imperative might be more fully realized. But this picture is not likely to happen. More likely is a situation in which "only the outlaws and police have guns." Whether such a picture is tenable or not depends upon the public will. Are people willing to adopt different risk control strategies in order to protect their safety? Since a certain amount of the firearm injuries occur within the context of law-abiding citizens themselves, there might be an immediate lowering of deaths and shootings. Perhaps, over time, only the well-connected criminals will be able to possess a gun. In that case, the death toll will drop even further. Wouldn't this be a safer society if such results were to come to pass?

If only a small portion of this vision were to be effected, then fewer emergency room surgeons might be faced with that gut wrenching picture of a young man or woman lying before her: a needless victim of the modern weapon with the highest damage coefficient: guns.

NOTES

[1] The most recent statistics relate 28,500 deaths due to firearms in the United States during 1998, *Health U.S. 2000* (Washington, D.C.: The National Center for Health Statistics, July, 2000).

[2] I should note here that 'ought' in this context has to do with an ethical or moral ought. This is viewed (as all ethical questions) in such a way that it will apply generally to all people in all morally relevant situations. The perspective and many examples of this essay concern themselves with the United States, but this is only because the author is choosing to historically situate the essay in this venue. The greater purpose will be to elucidate general principles that apply the use of guns in a society, per se. Thus, this essay will intentionally avoid all discussion of the United States Constitution and its guarantee of the right to bear arms. This is because if it is morally permissible for citizens generally to carry guns or not, the laws of the country should reflect this moral position and not the other way around. Thus, if a people (for example) do *not* have a moral right to bear arms, then the fact that the United States' Constitution says they do is morally irrelevant. Since this essay is about what should be the case (morally), it will ignore these other issues.

[3] *Crime in the United States: 1999* (Washington, DC: FBI, 2000).

[4] Obviously, there are silly counterexamples such as grenade missile launchers et al., but the point here is commonly possessed individual weapons. Thus bombs and lethal germ sprays would not count since they are not common or are not capable of being deployed by an individual without the support of some sort of war or terrorist structure behind them.

[5] For an argument detailing the Personal Worldview Imperative in relation to traditional theories of ethics, see: Michael Boylan *Basic Ethics* (Upper Saddle River, NJ: Prentice Hall, 2000), Introduction and Chapter Eight. This argument is enhanced by Michael Boylan and Kevin Brown in *Genetic Engineering* (Upper Saddle River, NJ: Prentice Hall, 2002), Chapters One and Two.

[6] For an exposition of the nature of this dialectical process see *Basic Ethics*, ch. 8.

[7] Those who aspire to be autonomous but do not do so from a sincere and authentic worldview are like the tyrant that Socrates logically shows to be flawed in his debate with Thrasymachus in *Republic* I, 338cff. Such a tyrant can make mistakes and hurt himself, may mistake his proper aim, overreach his own kind, or dilute his intent because he acts from a divided soul.

[8] I discuss the common body of knowledge in greater detail as it pertains to logical argument in *The Process of Argument* (Englewood Cliffs, NJ: Prentice Hall, 1988), chapter 1.

[9] I have mentioned healthcare (as a means to attaining health) as a basic good—thus incurring the correlative duty of society to provide the same in my essay, "A Universal Right to Healthcare" *International Journal of Politics and Ethics* 1.3 (2002): 197-212.

[10] This position has been advocated by Samuel C. Wheeler, Jr. "Self-Defense: Rights and Coerced Risk Acceptance" *Public Affairs Quarterly* 11 (1997): 431-443.

[11] There may be other arguments that might be made for these groups as well as other groups of ordinary citizens who might contend that they should possess guns. However, it is the contention of this essay that the above classification covers the major groups.

[12] 13.8% of the murders were among family members and 33.9% of the murders were among those who were well known to each other: *Crime in the United States: 1999* (Washington, D.C.: FBI, 2000).

[13] I believe that this is also the case with other weapons. I have anecdotally read in the newspaper about countless cases of women joggers (for example) who were assaulted, raped, and murdered even though they had mace and pepper spray concealed somewhere on their person. The reason, I suggest, is that the deterrent is useless unless one is forced to either practice the maneuver regularly or (better yet) is forced to actually confront it on a daily basis.

[14] Some might contend that my scenario does not take into account the situation of a man or woman who either live by themselves or who have no contact with children—actual or potential. There is a certain portion of the population to which this applies. In those cases the individuals could keep loaded guns in their bed stands. However, this does not diminish the prospects of accidental shootings (since the individual in this case might shoot before her head had sufficiently cleared). I think that this example reduces to the one mentioned earlier concerning concealed weapons.

[15] Hugh LaFollette has argued recently that instead of increasing safety, guns actually promote crime. He bases this conclusion upon an evaluation of statistical correlation. See: "Gun Control" *Ethics* 110.2 (2000): 263-281.

[16] However, the more common reaction is withdrawal and passivity. When the deck seems stacked against you, the principal virtue becomes endurance.

[17] The most recent factual source on this comes from: *Children and Families at Risk in Deteriorating Communities: Hearings Before the Subcommittee on Human Resources of the Committee on Ways and Means, House of Representatives, 103rd Congress, December 7, 1993.*

[18] When one adds to this the fact that often those living in these conditions do not deserve their fate, then this additional burden demands a response due to distributive justice. For a more detailed version of this argument see: Michael Boylan, "The Future of Affirmative Action" *Journal of Social Philosophy* 33.1 (Spring, 2002): 117-130.

[19] Some might reply, "What if everyone were forced to go two or more hours a week to a gun safety class so that everyone's skills were sharp and ready for action if need be?" To this I would say that my objec-

tion concerning weapons and personal safety (in some respects) might be satisfied, but at what cost? Wouldn't it be better if we have to mobilize the citizens on behalf of a cause to choose something like eliminating illiteracy or fighting poverty?

[20] I describe this test in more detail in *Basic Ethics,* Introduction.

[21] This sociological correlation has been effectively discussed by: Charles Green, ed. *Globalization and Survival in the Black Diaspora: The New Urban Challenges* (Albany, NY: 1997); Seldon Danziger, ed. and Ann Chih Lin, ed., *Coping with Poverty: The Social Contexts of Neighborhood, Work, and Family in the African American Community* (Ann Arbor, MI: University of Michigan Press, 2000); and Jorge J.E. Garcia, and Pablo De Greiff, eds. *Hispanics/Latinos in the United States: Ethnicity, Race, and Rights* (London: Routledge, 2000).

[22] A lesser differential exists among African American women, 5.0 per 100,000 as opposed to 3.0 among white American women. However this is still a shocking differential. The statistics represent 1998 (the most recent year cited), *National Vital Statistics, July, 2000* (Washington, D.C.: National Vital Statistics, 2000).

WANDA TEAYS

FROM FEAR TO ETERNITY:
VIOLENCE AND PUBLIC HEALTH

ABSTRACT: Interpersonal violence is a widespread social problem. As an ethical concern, it has generally escaped the attention of bioethicists and philosophers. This needs to change for reasons set out in this
paper. Violence is gendered: victims of gun-related violence are overwhelmingly male; those of all other
crimes are overwhelmingly female. The care and treatment of these patients deserves careful study. By
assuming the frame of reference of violence survivors, we gain greater insight into what they are up
against. This involves an examination of the key existential and institutional barriers that can inhibit or
slow the patient's recovery. Existential barriers are personal, psychological, spiritual, or ontological; they
are the internal obstacles violence victims face. Institutional barriers are socially-constructed obstacles
these patients are confronted with. An awareness of both sets of obstacles is necessary to effect change in
this area.

KEY WORDS: Violence, Domestic violence, Female victims of violence, violence survivors

Fear entered me.
It always inhabits me.
My dreams today are still
Filled with these frights.
—*Georges Jeanclos*

I lived with sudden fear
The way others live with cancer
—*Nancy Raine*

Avoidance is the order of the day when it comes to violence. As a public health concern, interpersonal violence is a daily reality—yet it has received little attention by
bioethicists and philosophers. That can—and ought to—change. We can make a difference in the lives of the survivors, their family, friends, doctors, nurses, and others
drawn into the circle of recovery and healing. This is a call for bioethicists and health
caregivers to reflect more deeply on the toll violence has exacted and to become more
proactive about addressing this moral problem.

We need to examine the existential and institutional obstacles violence survivors
face in navigating the health care system. In this way, we will be better able to help
both the patients and caregivers whose lives have been touched—and, for some,
changed forever—by gunshots, stabbings, assaults, rapes, beatings, incest, elder

135

M. Boylan (ed.), Public Health Policy and Ethics, 135-165.
© *2004 Kluwer Academic Publishers. Printed in the Netherlands.*

abuse, stalking, or other acts of violence. Existential obstacles are the personal, psychological, spiritual and ontological barriers that a victim of violence will face. For some they are inner demons; for others, they pose much less imposing hurdles. Institutional obstacles are the external, socially-constructed barriers that a victim of violence confronts when dealing with hospitals, clinics, and other establishments—as well as the attending personnel and general public. For some, they are oppressive, prejudicial, or discriminatory forces; for others they act as unwarranted assumptions to be identified and cast aside.

Far more needs to be done to prevent violence—but this lies outside of the scope of this paper. I believe we must first focus on violence survivors. If we can clarify what victims of violence are dealing with, we might then arrive at a more empathic public healthcare response. This requires a foundation of sensitivity and respect, as well as strong channels of communication and a patient-centered support system.

By assuming the frame of reference of violence survivors, by listening to their concerns, we can acquire insights unattainable by other means. In this way, we will be able to achieve a more nuanced ethical analysis of violence, formulate guidelines, and lay the moral groundwork to effect social change. I start by looking at the range and extent of interpersonal violence, in part one. As the data reveals, there are significant issues of asymmetry in terms of gender across the various categories of violence victims. This calls us to recognize the human cost of those directly or indirectly affected by violence and suggests areas for further research and critical inquiry. In parts two and three I turn to the personal and societal impact of violence, examine the obstacles, and offer some recommendations. Let us then start with the big picture.

PART ONE: VIOLENCE CASTS A LONG SHADOW

Most of us have witnessed the ravages of violence on friends or family. Most of us have seen the ways in which violence can affect whole communities, as when children are kidnapped, bystanders killed in random shootings, neighbors beaten in robberies, teenagers scarred by gang violence, and so on. That this has captured the attention of few ethicists doesn't diminish its significance. "Too often," remarks Rebecca Dresser, "bioethicists focus on the micro-problems, leaving the larger, systemic ones untouched."[1] Certainly, violence touches many lives, often to a devastating effect.

Dealing with violence is not like anything else in health care. The course of decision-making differs from that of treating a disease. And yet some types of violence resemble a medical epidemic and victims have been known to compare their trauma to getting cancer. Unfortunately, violence lacks the cachet associated with other areas of medicine and tends to be relegated to the margins. For one thing, it's messy—not just in real life, but conceptually as well. For example, it gives rise to issues of autonomy, but does not normally involve informed consent. It has both personal and societal repercussions. And the very people involved in treatment or research may have been victims or perpetrators at some time in their own lives.

This creates a divergence from most medical issues. Generally, health care concerns and treatment options do not involve law enforcement personnel, lawyers, social workers, counselors and the like. Few other medical issues, aside from contagious disease, put doctors and other personnel at risk. Interpersonal violence raises a unique set of concerns that merit serious consideration. Acknowledging the obstacles facing victims of violence is crucial. Otherwise, our theories, policies, or treatment plans will be based more on speculation than knowledge. But if we approach the issue from the frame of reference of the victim, the benefits will be considerable.

One of the sticky issues is the role individual autonomy plays in the choices and response of both perpetrators and victims of violence. In his essay on gun-related violence Leigh Turner said, "I am disturbed that the very ethicists who place such an emphasis on the principle of personal autonomy neglect the many ways in which autonomy, more thoughtfully considered, can be impeded."[2] That's the rub. Violence impedes the actions, the thought, and the existential reality of all it touches. Like a medical epidemic—such as the outbreak of Severe Acute Respiratory Syndrome (SARS)—it gives rise to broad social and political issues. And, like an epidemic, it warrants global, as well as individual, assessment.

Virginia Warren recommends that we look at empowerment alongside the concept of autonomy, which—as Turner suggests—is a broader issue than typically dealt with. Warren argues that empowerment "calls for changes that are more radical, political, and creative than does standard autonomy."[3] How we define and discuss autonomy has implications for policies and practices in treating victims of violence. Both physicians and patients need to assess the status quo: their insights can help us decide what needs be addressed and what factors should shape further discussion.

Examining the dark realm of violence, however, is a more likely pursuit of artists and filmmakers than ethicists and policy-makers. This is certainly understandable. It *is* unsettling, even repugnant, to realize that this patient wouldn't be there in front of that doctor were it not for the actions—the moral agency—of another. Most of us would rather not think about this at all. Yet it is ironic how many people seeking to be entertained by murder, rape, and all sorts of grisly violence in novels and movies close their eyes to the damaged lives around them. They may be surprised—even horrified—to learn how many victims and perpetrators of violence actually inhabit their world. Susan Brison suggests that, "The prevalent lack of empathy with trauma victims... [results] not merely from ignorance or indifference, but also from an active fear of identifying with those whose terrifying fate forces us to acknowledge that we are not in control of our own."[4]

This fact has not gone by unnoticed. The American Medical Association (AMA) acknowledges the avoidance of the topic by its members, noting the absence of national standards and the lack of preparation by physicians to deal with such acts of violence as sexual assaults. As the AMA reports:

> Because the medical and psychiatric treatment issues of sexual assault overlap, many physicians are *not prepared to* diagnose and to provide comprehensive treatment for the acute and long term effects of sexual assault. *Few medical school curricula address* the topic substantively in undergraduate or graduate medical education. Some hospitals and emergency departments have protocols in place to assist physicians in acute care settings, but there is *no national standard*. Primary care physicians face an even more complex problem, because many physicians and their patients are *unfamiliar and uncomfortable* with addressing patients' sexual history as part of traditional medical care. [5]

Similarly, the ethical dimensions of violence have received scant attention by bioethicists. Given at least some of those working within medical institutions are often overwhelmed by victims of violence, the relative silence on the topic is a cause for concern—if not outrage. But blame casts a wide net. In fairness to the medical profession, we haven't done a good job assessing the range and extent of violence in our society. We have yet to determine what ought to be the public response. With each well-publicized murder, school killing, child kidnapping, and the like, we are prodded to once again reflect on individual and collective responsibility—as well as prevention and treatment.

Liana B. Winett points out that only in the last 25 years have we seen the issue in terms of its relation to public health:

> One of the first times that violence was publicly recognized as a public health concern at the Federal level was in the 1979 Surgeon General's report... Violence was seen as an important contributor to morbidity and premature mortality. Later, in 1985, the Institute of Medicine published *Injury in America*, a treatise on the mounting threat that intentional and unintentional injury posed to public health. In October of that year, then-Surgeon General Koop convened the Surgeon General's Workshop on Violence and

Public Health, which cast a national spotlight on reconstructing violence as a public health concern. At the workshop, William Foege, then Assistant Surgeon General and Special Assistant for Policy Development at CDC, noted that until the early 1980s violence had typically been seen as a law enforcement or welfare problem.[6]

And so the spotlight widens beyond law enforcement to public health. That it is having an impact seems undeniable. We see this with gun-related violence, for example: It is reported that one in six pediatricians have treated a minor who is a gunshot victim and 77% of Americans believe that guns are a public health problem.[7] This should give us pause.

The broader focus also reveals the ripple effect of violence. Research indicates that medical practitioners may suffer indirect violence: "vicarious" or "secondary" traumatization and "compassion fatigue" show up in service providers assisting victims of violence. As noted by Sharon M. Wasco, Rebecca Campbell, and Marcia Clark:

[These caregivers] may experience indirect psychological effects of trauma, which can significantly alter the way victim-helpers perceive the world and have lasting impacts on their feelings, relationships, and lives. Astin (1990) reported that her work with rape victims resulted in nightmares, extreme tension, and feelings of irritability, lending support to models of indirect traumatization. Empirical studies have also documented similar experiences among mental health service providers.[8]

An ethical assessment of interpersonal violence would seem to benefit health care providers as well as patients. We need to acknowledge the concerns of those directly or indirectly affected by violence—the primary and secondary victims. "Practitioners themselves sometimes feel silenced as a result of what they know," asserts Barbara Sparks.[9] Consequently, the circle of impact is wider than previously imagined.

The prevalence of violence throughout all levels of society would startle us were it not taken for granted. From widely-reported acts of terrorism to more private incidences of gun-related shootings, gang violence, incest, sexual assault, date rape, domestic violence, elder abuse, stalking, and cyberstalking—violence takes many forms. The statistics might overwhelm, even numb us given the gravity of the problem. Just look at the statistics:

Profile by Gender: In the year 2000, more than 3.6 million males were victims of violence; 3% of these were caused by an intimate. More than 2.7 million females were victims of violence; 21% of these were caused by an intimate.[10]

Sexual Assault: In the year 2000, 260,950 rapes or sexual assaults were reported. Of these, 14,770 or 5.7% were male victims and 246,180 or 94.3% were female victims.[11]

Sexual Assault of Children: 61 percent of female assault victims are under the age of 18 and 20% are under the age of 12. 43% of those are 6 and younger. 33 percent of girls and 16% of boys are sexually abused before the age of 18.[12]

Stalking: As of May, 2001, one out of every 12 women (8.2 million) in the United States and 1 out of every 45 men (2 million) have been stalked at some time in their lives. One percent of all women (984,000) and .4% of all men (36,000) had been stalked during the 12 months preceding the 2001 survey.[13] Women are twice as likely as men to be stalked by strangers and eight times as likely to be stalked by former intimates.[14]

Gunshot wounds: In 1998, 32,436 people died from gunshot injuries and an estimated 97,308 people went emergency rooms with non-fatal gunshot wounds.[15] 85 percent of all gun fatalities involving young victims are males (a 5.5 to 1 disparity with females), and the gun-related fatality rate for black males is 2.4 times as high as that for Hispanic males, and 15.3 times as high as that for non-Hispanic white males. [16]

Domestic Violence: Battered women account for approximately 22% to 35% of women seeking care for any reason in emergency rooms and 23% of pregnant women seeking prenatal care.[17] As reported in 2002, 39% of women studied had experienced one or more episodes of violence by their partner. Of these: 25% had experienced two or three types of violence, 20% four to seven types, and 31% eight or more types.[18]

Elder Abuse: In 1999, 4 per 1,000 people over 65 years of age were victims of violence.[19]

The victim profile shows that violence is gendered:
- victims of gun-related violence are overwhelmingly male
- victims of stalking, sexual assault, and domestic violence are overwhelmingly female
- women are about 6 times more likely as men to be victims of violence by an intimate partner[20]

These discrepancies—the asymmetry in the different categories—warrants attention if we seek the connections between acts of violence and societal attitudes or values.

Nevertheless, some generalities about violence are gender-blind. You don't have to be a female to have fear consume you. And the need for validation and affirmation is as real for males as it is for females. Similarly, the need of victims to pick up the pieces of their lives, to take control over their affairs, and assert themselves with dignity has little to do with gender. I have no reason to think concerns raised by female violence survivors do not apply equally to males. However, more research needs to be done in this area. The uniquely different situations and gender-related struggles should be brought to light. That work lies ahead.

The personal, social, political, and public health aspects of violence warrant extensive study and a thoughtful response. Social justice requires us to look more closely at the epidemic of violence and its ethical ramifications for public policy. It is, not too late for us to turn this around. Rosemarie Tong recommends a cooperative effort in addressing ethical problems:

> [E]thics, as well as bioethics and public policy, is about doing good as well as avoiding evil. As I see it, ethics is not only about the rights of individuals and their quest to develop themselves, but also about the responsibilities of individuals and their desire to relate to each other. Ethics encompasses more than just the rules that enable individuals to pursue their separate interests; it also comprises the ideals that enable individuals to cooperate toward the achievement of common purposes.[21]

Such a collaborative effort could yield many benefits and help us create a system that can more ably facilitate healing and recovery than is currently in place. The breadth of a team approach with shared goals and a commitment to individual dignity would have considerable value here.

One systemic issue is how to address the needs of both patients and health caregivers. Emergency departments, hospitals, doctors' offices, and clinics face issues about both policy and practice concerning acts of violence. Aside from the impact on violence survivors, the reality of health caregivers must at times seem Dantesque—downright hellish. The sheer horror of patients traumatized by violence cannot help but weigh on those providing medical treatment. Out of compassion for all concerned, we ought not avert our eyes any longer.

Let us start by asking if the process of healing and recovery is humane and caring. To answer this affirmatively, it helps to assume the option of the oppressed and examine the needs of victims of violence from *their* perspective. This will help us clarify both the public health aspects of violence and the steps necessary for systemic change. To do this well, we need to incorporate the victim's perspective into the dialogue. It will also give us a depth of understanding not as easily attained from a more detached perspective. This we will consider in the next section.

PART TWO: FROM FEAR TO ETERNITY—
THE VICTIM'S PERSPECTIVE

Victims of violence see their lives transformed. They have to draw upon whatever resources they can muster to confront their fears and get their lives back on track. Some have to reassemble the parts of their selves that were scattered, or shattered, by the trauma they suffered. As Susan Brison said of her assault and near-murder:

> A year after my assault, I was pleased to discover that I could go for fifteen minutes without thinking about my attack. Now I can go for hours at a stretch without a flashback. That's on a good day. On a bad day, I may still take to my bed with lead in my veins, unable to find one good reason to go on.[22]

Those victimized by violence generally have compound injuries—psychological as well as physical. However painful are the physical wounds, the component of fear makes the psychological terrain difficult to navigate. Fear wreaks havoc on the victim's state of mind and emotional equilibrium. This much is true: the end of the physical suffering does *not* herald the end of the fear, torment, nightmares, or sense of unreality transforming the lives of violence survivors. And there is no clear indication when, if ever, that fear will cease.

For the lucky ones, the fear that links the victim to the perpetrator subsides over time. Nevertheless, many victims' lives are fundamentally changed. Their metaphysical reality—their sense of their own personal identity, their perceptions of the world and relationships with others—does not automatically regain its pre-trauma shape. For some, nothing is ever the same again. Part of regaining one's dignity and self-determination is the reconstruction of a sense of normalcy—a personal status quo. It is a testimony to the human spirit that so many who have struggled with violence and post-traumatic stress have reassembled the parts of their lives and have had hope restored.

To develop an appropriate response from medical caregivers, we need to appreciate the challenges facing victims of violence. Elizabeth M. Schneider's remarks about the legal system apply to the health care system as well:

> In the expert testimony area, courts have recognized that the experiences of battered women are distinct and shared, that these experiences are outside the common experience of jurors, and that it is necessary for the jury to learn about these experiences in order to overcome myths and misconceptions concerning battered women in order to evaluate whether the woman was acting in self-defense.[23]

Understanding the major obstacles to recovery is required to arrive at an effective protocol and treatment guidelines for healthcare professionals. Without this knowledge, our methodology and procedures would have limited value.

The obstacles fall into two broad categories— those that are self-generated (existential obstacles) and those others place in the path of recovery (institutional obstacles). There may be other obstacles, but these are prominent:

A. Existential Obstacles
1. Fear
2. Distrust
3. Self-Blame
4. Diminished Self Worth
5. A Sense of Atemporality and Unreality

B. Institutional Obstacles
1. Disbelief
2. Denial

3. Blame
4. Misconceptions and Insensitivity
5. Poor Communication

An awareness of both sets of obstacles can help erect, or fortify, a more caring process for response and treatment of victims of violence. Let us first consider the existential obstacles facing these patients. Although physicians and others can play a supportive role, healing from violence requires victims to feel in control of their destiny.

Existential Obstacle #1: Fear

Of the existential obstacles, the most persistent is fear. For many, there is an all-consuming fear, fear without end, fear that takes hold like a vise grip. This is the first, most visceral, and generally the most long-lasting effect of violence. Fear often precedes, accompanies, and endures long after the traumatic event. Even threatened violence can elicit crippling fear. We need to see what this is like. We need also to remember that many acts of violence or threatened violence are part of a series with a beginning but no end in sight. In such cases, fear is amplified, stretched to eternity.

Imagine being the prey of a cyberstalker, getting anonymous death threats every time you open your email account. Personally addressed to you are one after another of email messages such as:

"I will curse you"
"I wish you were dead"
"I will kill you"
"I hate your mean face"
"I know your home address now. Finally!!!!"

Would it matter if your stalker did not threaten to kill you, but sent you hundreds, even thousands of emails containing testimonials of *love*? "One woman I know is getting 20,000 e-mails per day that say 'I love you'.... but there's no threat, so it's not a crime," says Colin Hatcher of SafetyEd, a company that helps victims of Internet stalking.[24]

Cyberstalking is no less threatening than physically stalking a victim; in fact it may be more terrifying. Although such a victim may not initially seek medical care, the fear, stress, and possibility of physical harm makes this area important to include in the discussion. In an article about a University of Michigan student who posted violent sexual fantasies about a fellow student, *The Chronicle of Higher Education* notes that,

Anonymous electronic messages may evoke a degree of fear that is not aroused by the spoken or written word, if only because the recipient may not know whether the sender is halfway around the world or in the next room. Especially in this relatively early phase of digital communication, a higher degree of apprehension may be warranted.[25]

Unfortunately, advances in technology have opened up channels to quickly strike fear into the heart of a victim. An article on email threats in the journal *Security* focuses on the impact this has had on the workplace, such as:

> Worldtalk's recently released Internet Email Corporate Usage Report indicates that 31 percent of e-mail exchanged with the Internet is a potential security threat. What's equally important to security executives is that emerging electronic communications threats, ranging from a spiteful voice mail to Web hacks, are increasing in number... the occurrences can happen to anyone and the harm can be extraordinarily significant. [A Midwest company manager, Gary Thompson] alleges he was the target of libelous voice mail spread by some unidentified person throughout the security executive's company... such incidents can happen to anyone anywhere...such occurrences can lead to tragic, complex results....[26]

Menacing phone calls and voice mail messages may be another source of intimidation or threats. According to a report in the *Journal of the American Medical Association*, "people who get repeated unwanted phone calls sometimes obtain an unlisted number. A stalker finds the new number in 48 hours on average, reinforcing the person's sense of helplessness."[27] The fear is palpable as you walk outside, answer the phone, or open another email—not knowing if your stalker is still there, in the shadows of your life. "It is now more than four years since I realized that a patient had begun to follow me," an anonymous victim writes for the *British Medical Journal*. The author continues:

> When activity from the stalker was frequent ...I found myself dipping into the realms of paranoia. I examined every car, studied every number plate, my stomach lurching if I glimpsed an abnormal loping gait... It is just as with people with cancer who hope that surgery has cured them—they cannot know for sure, just as their physical scars remind them of the trauma of surgery, pain in the operation site brings back fear of the disease, and follow up appointments reawaken their and their families' anxieties, so my "ailment" is frequently at the front of my mind.[28]

What is so disturbing, too, is that such fear is not easily dissipated. The flash of a stranger's face resembling that of the attacker can set back recovery to Square One. The smell of liquor on the batterer's breath can start in motion a chain of terror-laced apprehension. A friend's email with subject line "Hohoho" is too close to the "Heeheehee" death threat and, so, turns your insides to stone. What were once ordinary aspects of life now appear ominous; innocent practical jokes that used to make you laugh now set loose an inner reign of terror. You are overcome by fear that violence is just round the corner. You live with the fact that at any time, at any place, you could be killed.

Picture this going on in your life. Now picture it for the rest of your life. "My attack occurred over 20 years ago, and the legacy of that horrible afternoon has stayed with me, ...And I still worry that he'll find me and rape me again."[29] That was a female rape victim. Male rape victims suffer fear as well, as we see in the following description: "For years he would tremble uncontrollably, suffered panic attacks, had chronic insomnia, and attempted suicide."[30]

The ongoing distress can be palpable: "Effects of stalking may linger after the stalking stops," notes Lynne Lamberg in *JAMA*. These include "heightened fear and perception of vulnerability. People who have been stalked therefore may need treatment for depression and other symptoms of posttraumatic stress disorder"[31] Fear, depression, and post-traumatic stress—all are tied together by violence. It is because of this centrality of fear that I see it first on the list of existential obstacles. In constructing a public health care response to violence, therefore, we need to pay close attention to fear as an indicator. For instance, women who stated that they were sometimes or often afraid were 32 times more likely to have experienced violence than those who reported that they were never afraid; fear of partner and experiencing controlling behavior were significantly associated with domestic violence; and anxiety is more strongly associated with domestic violence than depression."[32]

It would be beneficial if we more explicitly affirmed the patients' attempts to wrestle down the fear and anxiety so much a part of their lives. As a resident of a high-crime neighborhood said about the debilitating fear of being shot: "I've got to get over my fear. It controls you. It does not allow you to be. It makes you feel like a prisoner when you have not committed a crime."[33] Feeling that one's freedom has been curtailed by fear generates a back draft of resentment or depression that health caregivers need to be on the watch for when treating victims of violence.

Existential Obstacle #2: Distrust

The second existential obstacle, distrust, can also have a long-lasting effect. Distrust unchecked can lead to paranoia. Victims often feel cut off from others, consigned to a realm of solitude. Take the case of Angela Moubray, who enjoyed Internet chats on wrestling, soap operas, and the like. Then, one day, a fellow participant began sending menacing email:

> Soon, she says, he barraged her with a stream of threats such as "I hope you get raped." Over nearly two years, the Virginia resident received unrelenting messages from a person whom she had never met, culminating in the missive: "I will kill you Ang, I mean it."...the warnings she received terrorized her so much that she had to take safety into her own hands. "I started carrying pepper spray, and I wouldn't go anywhere alone. My Dad bought me a gun," she says.[34]

We may find it hard to comprehend what Moubray's life was like. But it seems safe to say that it underwent a radical transformation. The terror she experienced was isolating and made her suspicious of others. David Winthrop Hanson comments on this phenomenon with respect to battered women. He observes that, "even if the woman chooses to leave the confines of her home, the actual prison in which she lives has no boundaries, but rather surrounds her continuously."[35] Trust is one of the first casualties of violence. The link to what was once the norm has been severed.

Susan Brison describes this experience: "Unlike survivors of wars or earthquakes, who inhabit a common shattered world, rape victims face the cataclysmic destruction of their world alone, surrounded by people who find it hard to understand what's so distressing."[36] Another violence survivor, Nancy Raine, echoes this point: "The attacks caused me to be cut off from everyone to whom I should have been close. I pretended to be 'normal,' acting from cue cards that flashed in some part of my mind. I could not share my terror because I did have a remnant of sanity—I knew it was crazy."[37] To create a more caring health care response, we need to acknowledge how powerful the sense of isolation can be.

Existential Obstacle #3: Self-Blame

Isolation and distrust often eat at violence victims, causing them to ask themselves, "Why me?" This can give rise to the third existential obstacle—self-blame. We live in a society that is conflicted in terms of assigning blame; thus it is not surprising that victims turn on themselves. This may arise from a desire to have some control over what happens to us. Lynne Lamberg's remarks are applicable here; namely:

People who are stalked often drastically alter many aspects of their daily lives. ...Some develop anxiety disorders and increase their drinking and smoking. One in three is physically and/or sexually assaulted by the stalker. One in four ruminates about or attempts suicide. A few have been murdered. Despite these grim costs, *such people often believe they must have done something to provoke the stalker,* and do not tell others...[38] (emphasis mine)

For many, self-blame and self-imposed silence form the default position when confronted with acts of violence. This may point to systemic problems—such as misogyny, racism, or homophobia. "It can be less painful to believe you did something blameworthy," Brison argues, "than to think you live in a world in which you can be attacked at any time, in any place, simply because you are a woman."[39]

Violence against women *does* seem linked to the devaluation of women. But the effects of oppression don't end there. For instance, Black and Latino males may not find their gender any protection against gun-related assaults. And poverty is more likely to put you at risk of violence than wealth.

Although women are the most prevalent victims of non-gun related violence—and we certainly need to give this our utmost attention—there are other concerns that are also pressing. We need to go beyond gender to a broader notion of power dynamics and justice issues. Violence does not take place in a social vacuum. To assess its impact on public health, we need to examine the extent of prejudice, discrimination, and injustice woven into our cultural fabric. The medical establishment is not untarnished by such concerns.

Whatever oppressive societal forces may be at work, self-blame attempts to bring order to chaos. The assumption that we must have done something wrong to warrant such "punishment" restores balance, however tenuous it may seem. In some cases—particularly with intimate relationships—this leads the victim to forgive or excuse the perpetrator. Accepting blame then allows the victim to retain a semblance of self-respect and autonomy. Clearly, this is a risky path to go down. As Sandra Lee Bartky asserts, "the consciousness of victimization is a divided consciousness."[40]

If self-blame is a form of self-oppression, it must be overcome to re-establish a sense of wholeness. The divided consciousness can then be mended. At that point, blaming oneself can be seen as the obstacle that it is. "My repudiation of the victim identity," bell hooks observes, "emerged out of my awareness of the way in which thinking of oneself as a victim could be disempowering and disenabling." She holds that "to name oneself a victim is to deny agency."[41]

Within the medical establishment this should signal a change in protocol.

Knowing that victims of violence struggle with issues of self-blame, doctors and other caregivers can help patients put their experience into words and shape a narrative. More openness may help create a more constructive and compassionate response on the part of the medical establishment. Feeling they are in a safe environment where the health care team is a positive, caring, and non-judgmental presence allows patients to put down their guard—even just for a few moments.

Existential Obstacle #4: Diminished Self-Worth
Care must also be taken to keep things in perspective. For many victims—as in the case of stalking and domestic violence—the violence is not perceived as *over* simply because they are getting medical treatment or there is a cessation of violent episodes. This is not always easy to express. Emotional distress, confusion, panic or fear can lead the patient to feel that no environment is truly safe and restrict the ability to communicate. According to Abby Wilkerson, this sometimes leads to self-imposed silence, as the patient struggles to come to terms with the situation.

> Unfortunately, a distinguishing feature of domestic violence is that its victims may not be in a position to explain the cause of their injuries. Debbie Burghaus, a social worker in a Chicago hospital, states, "a lot of times, the victim isn't going to offer to tell you she was beaten up, because he's waiting for her in the hall, or she's just not empowered to leave him yet"… Physicians "see their job as fixing the physical problem, and they don't see the problem, so that's not their job."[42]

Just being there, taking the time, and providing a caring presence helps establish trust between physician and patient. If this can be accomplished, the patient can more easily face and overcome the fourth obstacle: diminished self-worth. In the aftermath of violence, patients often struggle to appear normal, unscathed—all the while fearing rejection by others. We need to affirm the patient's vulnerability on this score. For example, in the case of domestic violence:

> The psychological abuse women experience can be as devastating, if not more so, than the physical violence. Although research is limited with respect to its independent effects on women's health, there is some evidence to suggest that psychological abuse lowers self-esteem and increases symptoms of depression and posttraumatic stress disorder among battered women.[43]

This debilitating effect—lowered self-esteem—warrants attention by caregivers. Although sexual assault victims may more visibly demonstrate this problem, watching for such signs should be included in the public health response to all kinds of violent episodes. Susan Brison's work on the ontological consequences of violence may help ethicists as well as medical caregivers assess the significance of this obstacle.

Existential Obstacle #5: A Sense of Atemporality and Unreality
In terms of recovery, a sense of atemporality and unreality is another common existential obstacle. Freud may have been right when he said there is no sense of time in the Id; that traumatic events occurring years, even decades, ago are but a millisecond away as far as the psyche is concerned. We see this with violence victims, who often act as if time has stood still. Perhaps this is a survival mechanism. For example, in his study of combat trauma in Vietnam veterans, Jonathan Shay claims that, "The destruction of time is an inner survival skill."[44] Auschwitz survivor Charlotte Delbo also sees a link between trauma and a sense of unreality: "Was I alive to have an afterwards, to know what afterwards meant?" she asks. She then observes, "I was floating in a present devoid of reality."[45]

One issue for the public health system is that those traumatized by violence cannot be expected to just "let go" of the past, "snap out of it" or "don't look back." *Past, present,* and *future* constitute a triad no longer stable in a victim's consciousness. The trauma of the present violence may block, or severe, access to the past. And a future beyond the present overwhelmed and overloaded by a sense of unreality may simply seem unattainable. Consequently, one task facing doctors and other medical personnel is to help in the temporal reconstruction of the patient's life.

Unless a sense of temporality and reality is restored, violence survivors risk being locked in an eternal present. This would leave them struggling to make sense of what happened while trying to prevent self-blame and self-recrimination from taking root. Otherwise, patients may be pulled into a downward, debilitating spiral. Reconnecting to a sense of past, present, and future provides a key framework for putting trauma into perspective and reestablishing a sense of hope. This is an important asset to recovery.

Here's where a caring, healing environment and a supportive healthcare team can make all the difference. For instance, "In many cases, a physician or nurse may be the only person women feel comfortable talking to about their partner's violence. This provides health care providers with a unique opportunity to identify and assist domestic violence survivors."[46] By acknowledging the existential obstacles patients are struggling to overcome, medical caregivers can act as a healing presence. They can be there as a part of the patient's support network. Just being there, providing a source of empathy and affirmation should not be underestimated.

The empathy model of *acknowledge-listen-affirm* can go a long way in assisting the violence survivor through the recovery process. As we have seen, there are significant personal obstacles that must be confronted. That is not the only thing at stake, however. We need also to recognize the other obstacles—the social ones—that face the victim of violence.

This may require us to look in the societal mirror, to reflect on the ways institutions and individuals within those institutions either help or hinder those who are seeking to navigate the system. At this point we need to grasp what the victim is up against in dealing with *others* along the path of recovery.

PART THREE: INSTITUTIONAL OBSTACLES IN THE HEALING JOURNEY

It is nothing short of amazing to see people who have suffered even monstrous violent crimes overcome their inner obstacles. The result may seem like a rebirth—so dramatic is the division between their lives before and after the violent incident (or series of incidents). But the hurdles they face are societal as well as existential. And this is one area that the medical establishment can take control and set change in motion.

To do so effectively, attention must be paid to societal and institutional factors, as well as the treatment options. In this regard, Mary Mahowald's advice may help. She says,

> Medical crises that occur because of gunshot wounds, rape, wife battering, drug addiction, and child abuse require treatment beyond that administered to the individual. From the vantage point of preventive medicine, what Friedman calls "care to the public domain" is crucial to the long-term recovery and maintenance of individual as well as social health. Health professionals have a special responsibility to exercise this kind of care by supporting societal change and programs that promote human health.[47]

Given the patient has already suffered physical and psychological harm requiring medical attention, it is vital that we help ease the recovery process. Removing institutional obstacles should, therefore, be a commitment of the public health system.

Institutional Obstacle #1: Disbelief

The first institutional obstacle is disbelief. The flip side of disbelief is validation. They are as different as night and day. All too often, victims have complained that others responded to them in uncaring ways that project an attitude of disbelief or incredulity. In recognizing the importance of validating the victim, the AMA set out strategies for dealing with victims of sexual assault. In *Strategies for the Treatment and Prevention of Sexual Assault*, the AMA notes that, like victims of sexual assault and domestic violence, stalking victims sometimes spend an inordinate amount of time attempting to convince others that they are being stalked and are in real danger. According to the AMA, victims deal with varying reactions from others, including:

- Disbelief
- Blaming the victim for causing the stalking, particularly in cases where victims know their stalkers
- Believing the victim but refusing to help or support the victim
- Believing the victim, taking the victim's concerns seriously, and offering support and assistance[48]

The AMA goes on to recommend that service providers take the following steps when treating victims of violence:

- Validate that the stalking experienced by the victim is indeed occurring
- Understand the complexity of the crime and the danger posed by stalkers
- Have the capacity to provide effective intervention strategies that protect and support victims and restrict stalkers[49]

The significance of these points should not be minimized. It is abominable that victims should have to invest so much time and effort convincing others that they have endured violence—and may risk further harm. As the AMA's work on strategies attests, victims of stalking or other types of violence need to be taken seriously when they seek help. That many fear for their lives should not be treated as an overreaction or as delusional.

The evidence is already in that such fears are rarely exaggerated. Look, for example, at the number of women killed while restraining orders are in place. Keith E. Davis, April Ace, and Michelle Andra discuss issues around stalking and note that the potential for general violence and lethal violence has also received particular attention—whereby the simple situational factor of stalking is a risk factor for further assaults "just by virtue of the tenacious proximity seeking toward the victim, and especially if it occurs in combination with several other high risk behaviours."[50] They cite research indicating that in the 12 months before an intimate partner murdered (or attempted to murder) their female partner, more than 75% of the women had been stalked. They then conclude that both intimate partner assault and stalking are risk factors for lethal and near-lethal violence for women, especially when these two perpetrator behaviors occur together.[51]

It is vital that those responding to those subjected to violence or threat of violence not act in dismissive ways. Validation, not disbelief, is called for in such situations. It is callous to think patients traumatized by violence actually have ready access to police or other protective services. It is not true that restraining orders always work and that escaping violent relationships just takes determination. There is more ignorance than insight behind such attitudes. Anyone one seeking to address the public health system has to look at the inhibiting effects of disbelief on the part of doctors and other members of the response team. As a result, we need to look more carefully at societal beliefs and values around causes and effects of violence and the view that victims can easily extricate themselves from harmful situations.

Institutional Obstacle #2: Denial

Closely correlated with disbelief is denial and trivialization—the second institutional obstacle. Trivialization compounds the sense of helplessness and isolation. It does *not* improve the situation, or make it any less frightening for the victim of violence. However well intended they may be, attempts to make light of violence are ill advised. At best such attempts to humor the victim or trivialize the experience are futile and alienating.

Picture, for example, the victim of a stalker who is asked by a co-worker, "Have you heard any more from your *ghost?*" The "ghost" is all too real for the victim. Denying the seriousness of the situation just makes things worse. Writing about the problem of denial with respect to battered women, Michael Dowd states:

> Even the community of decent people finds society's inability to quell the tide of violence against women too horrific to accept. The reality of life for a battered woman is indeed beyond the knowledge of the average person, and while denying the truth of the battered woman's story can be easier than confronting it, the enlightened must aim to recognize the truth in order to foster society's acceptance of responsibility for the violence among us.[52]

He argues that recognizing the truth is a necessary step in societal—and I will add, institutional—transformation. To create a more credible response to victims of violent crime entails removing our societal blinders. We need to see the situation as clearly and honestly as we can. Otherwise, our blindness, denial, disbelief, and ignorance will likely perpetuate ineffective policies and procedures. This, in turn, forces violence victims back on their own resources. Rebecca Dresser reports that there are widespread issues in this regard:

Women continue to report that doctors and other health care professionals have "not listened to them or believed what they said; withheld knowledge; lied to them; treated them without their consent; not warned them of risks and negative effects of treatments"—the list goes on and on. Extreme discontent with contemporary health care pervades the women's stories in these volumes.[53]

Institutional Obstacle #3: Blame

One problem associated with inadequate listening skills is the misplacement of blame when patients seek help from medical professionals. That such blame and shame leaks in from the greater society is inexcusable. This third institutional obstacle needs to be acknowledged and addressed. Furthermore, the extent to which societal prejudice plays a part in blaming the victim merits further study.

We should take steps to minimize destructive prejudices in public health practices. From the victim's point of view those negative reactions can weigh heavily, especially when they come from doctors and other caregivers. Furthermore, accusations that they bear responsibility for what happened can exacerbate feelings of guilt. It seems unjust, given victims of such crimes as robberies, carjackings, and drive-by shootings are rarely thought responsible for their misfortune. Gail Abarbanel points out the asymmetry in such judgments:

> It's shame that your vulnerability has been exposed, that you failed to defend yourself, to prevent it from happening. There's a burden to the violation that doesn't exist for other crimes. ...[T]here's a reason people are more likely to say they were the victim of a robbery than to mention in conversation that they were raped. That's because of the response and reaction when they do tell. You don't get the immediate, unconditional support. People raise their eyebrows, they wonder, "What did she do? What could she have done?" It's the only crime where the victim still feels on trial.[54]

There seems to be a human need to ascribe responsibility for events in causal terms. The argument would then proceed as follows: For any given effect, there must have been a discernable cause. This is a metaphysical concept many consider reasonable and applicable to our lives. By this reasoning, then, any particular violent event has a specific cause (or causes). That the cause might be random or unknowable is unacceptable in such an explanatory schema. *That* perpetrator did not surely act on *this* victim without some reason. There had to have been a prime mover setting things in motion. Consequently, violence cannot logically arise without cause—which means victims are either catalysts or aggravating factors in the violence they experienced. We simply find it hard to comprehend that a victim of violence is without blame—unless the person was a child or incompetent.

Rationally we know this is absurd. Things happen every day without ascertainable causes. Nevertheless, the universe is much more tidy—and less frightening—if victims can be said to have a part in the causal chain of events. A corollary is that people are rarely completely innocent. Thus, logic and karmic morality point to the same conclusion: victims of violence must have done *something* to bring this sorry state of affairs on themselves. As we all know, there are two sides to every story. That's what we're told, anyway.

It is just such logic that we need to address. It is insidious and misguided. Health caregivers need to realize—both conceptually and ethically—that victims of violence may have done *nothing whatsoever* to deserve what happened. It is vital that any moralizing behind the assigning of blame be purged from our public health system. And the fact that perpetrators assign blame—e.g., the woman who burnt the eggs or talked on the phone too long is the reason the batterer smashed her face—does not legitimize their doing so. Nor should we assume a perpetrator's actions require provocation.

We must also realize when looking at moralizing about violence victims and perpetrators that neither may have admirable qualities. In other words, not all victims are sympathetic and not all perpetrators are monsters. People have been known to punch, kick, or goad one another, causing violence to escalate. There are also situations, like in gang violence, where each participant may be simultaneously victim and perpetrator of violence. It may be that neither victim nor perpetrator is morally blameless. We still have a health problem and the emotional issues underneath. Action must be taken without regard for health care professionals trying to make moral judgments on the patients.

Institutional Obstacle #4: Misconceptions
A related institutional obstacle facing patients are societal misconceptions around crimes of violence. Unenlightened—even Neanderthal—attitudes are as present in the legal and criminal justice system as in health care. No doubt they reinforce one another, creating a host of difficulties for the traumatized patient. Let's look at two examples to clarify this point.

First, both males and females can be raped, but male rape victims are an invisible minority. The AMA has taken steps to rectify the misconception that only females can be raped:

> Historically, the legal crime of rape and the notion of a legal age of consent
> for sexual relations was applied strictly to women... Furthermore, "rape" was

a legal term that applied only to vaginal penetration of a woman against her will and without her consent. Anal penetration of a man was referred to as "copulation with a member of the same sex or with an animal."[55]

When the term "rape" was only applied to females, male victims of sexual assaults commonly faced skepticism and insensitivity from those in the medical establishment. Fortunately, this has changed, although much more needs to be done in this area. We also see societal misconceptions in cases involving battered women. For instance, the court ruled in the 1981 decision of *People v. Powell* that "a battered woman does not like the beatings but *likes the loving behavior which occurs after the beating* and she becomes submissive and passive."[56] Of course, it's not that loving behavior is undesirable; the problem is in thinking that any such loving behavior compensates for the beatings. In addition, societal views about what counts as an "imminent" danger has significant repercussions for victims who kill an abuser. There is considerable debate as to whether they can use the justification of "self defense" in such a case.

Institutional Obstacle #5: Poor Communication
Fortunately, there are steps we can take. We can look more carefully at what the victims are up against. We can address the institutional hindrances they face. At that point, a trusting, open dialogue between the patients and their caregivers will be easier to achieve. Well-developed channels of communication are vital for all elements of public health—and certainly in the area of interpersonal violence, where things can be so unstable, and even volatile. Poor communication is a serious hindrance, which is why I listed it as our last existential/institutional obstacle.

Caregivers need to establish a relationship of trust and demonstrate patience in the recovery process. This needs to be clearly communicated to the patient. A lack of clear communication—one based on honesty and respect—can be a serious impediment in working through trauma. Those who have ever worked with war veterans suffering from combat fatigue know, veterans often go over the same material over and over, telling the stories, claiming some pieces, while purging others. This is true of most other victims of violence—the need for a narrative structure to tell and retell their stories, to communicate what they've been through, is fundamentally important to their recovery. Having a supportive presence in that process can be very healing.

Some think the society is moving forward in this regard. "Increasingly, women are unwilling to be silent victims," observes Linda Gross.[57] In the summer of 2002 there was a widespread debate over two teenage girls appearing on national television after their abduction, rape, and near murder. Many thought their willingness to go public showed progress. "In an ideal world," says Nancy Raine "all survivors would love it if this was a crime like all other crimes, where if it came up that you'd been raped, it would not plunge the listener into silence and discomfort."[58]

By refusing to be silent, victims may help dispel listeners' discomfort and with a more empathic response, patients traumatized by violence can regain some of their dignity. More openness, validation, and better communication go a long way to move violence survivors away from self-blame and self-victimization. However, we need to respect for the right of victims to decide how much to publicly disclose. We should strive to reach a balance between open communication and respecting an individual's privacy. We can work to achieve a more healing environment—one that validates rather than judges. The value for all concerned should not be underestimated.

RECOMMENDATIONS

Improved communication is crucial, but it should be handled carefully. As noted by the AMA, sexual assault, and presumably rape, often disrupts cognitive perceptions, rendering memory of events incomplete or inaccurate. This raises some concerns about how best to respond to patients who were assaulted. "Well-meaning but detailed discussion of the assault may further traumatize the patient," the AMA reports. Moreover, it is "erroneous to suppose that if the patient is encouraged to verbalize events and 'work the attack through,' they will feel better and be calmer. Sexual assault is a circumstance of such magnitude that it does not respond during the acute phase to emotional release or catharsis."[59] This may seem counterintuitive, but it points to the importance of patience on the part of all concerned.

To a great degree, these observations apply to other violence victims as well. On one hand, patients may desire to talk about the experience and put it into some comprehensible framework. On the other hand, intrusive or disrespectful questions or a hurried manner on the part of the medical caregiver may worsen the situation and make victims leery of such "help." We should strive for a model of *listening*, not a line of inquiry, when dealing with victims of violence.

Doctors and other caregivers may be reluctant to question patients when violence is suspected. They may fear intruding on the individual's sense of privacy or fear that openly discussing the violent episode(s) will further traumatize the patient. Research suggests, however, that we should work for more open and compassionate channels of communication. This may require physicians to ask questions and inquire about warning signs. Of course, such questioning needs to be handled tactfully and with sensitivity to the patient's sense of vulnerability.

Patients are more receptive to routine questioning about violence than doctors seem to realize or have put into practice. Talking about our experiences—good and bad—help us achieve order and meaning in our lives. A domestic violence study reported in the *British Medical Journal* revealed that 77% of all women patients were in favor of routine questioning about the issue by their usual general practitioner.[60] This openness to questioning on the part of physicians is also found in adolescents, as the following report on gun-violence indicates:

> Approximately two thirds of the students reported that they would confide in their physicians... If directly asked by their physician about guns in general or guns in the home, the majority of teens would be willing to talk about the issue... Also, half of those who have considered obtaining a firearm and half of those who have used a firearm in a threatening manner said that if asked, they would discuss the issue with their physician.[61]

One potentially problematic area of communication involves issues around the privacy rights of the patient, especially when the health care worker is dealing with a minor. More needs to be done in this area before we can rest assured that the system in place is working as well as it should. For example, in 1998 Congressman Don Manzullo of Illinois proposed a bill to the House of Representatives (HR 3230) that sought to make notification of parents or legal guardians a requirement of clinics seeking Title X funds before minors could obtain birth control.

He cited the case of a 13-year old girl who was being sexually abused by her teacher. Evidently, the teacher knew that federal Title X rules prohibited clinics from notifying parents when issuing birth control drugs to minors. The county health clinic gave the girl shots of the long-term birth control Depo-provera over an 18-month period before the sexual abuse was brought to light and the teacher arrested.[62]

One question raised by this case is how to resolve potential conflicts between the patient's right to privacy and other concerns—such as obligations on the part of health caregivers to disclose suspected abuse, or parental rights to oversee the care and treatment of their children. There are a number of relevant concerns here that have to be evaluated. With regard to the issue of communication, we want to encourage doctors and other medical caregivers to develop channels of communication and policy guidelines that would allow appropriate discussions, even questions, when they are facing such difficult cases as minors under the age of consent seeking birth control. Presumed privacy rights need not negate further inquiry when a red flag has been raised—whether by a sexually active child, a patient who has suspicious bruising, or the like.

Martha Minow's work on the public/private distinction is applicable to cases of domestic violence, incest, and child sexual abuse. In likening the home to a man's castle, for instance, laws and practices seek to preserve such "private" domains. This has had grievous consequences for women and children. Minow believes the policy of "nonintervention" by the state has bolstered the authority of the man. "State-created rules about what counts as a criminal assault," she says, "regulate the family, determining, for example, whether the state recognizes or refuses to recognize marital rape as 'real rape,' which can trigger the criminal justice enforcement apparatus."[63]

It is time to rethink boundaries and definitions that carry such significant consequences, particularly when the norm has shielded perpetrators. We need to remind ourselves that definitions, policies, guidelines, categories do not fall from the sky—they are the products of human activity. As Minow says, "The more powerful we are, the less we may be able to see that the world coincides with our view precisely because we shaped it in accordance with our view."[64] By consciously trying to understand the perspective of the other—in this case the victim of violence, we broaden our worldview.

That patients are willing to confide in physicians speaks volumes, for it reveals a basic level of trust already in place. In building an ethical framework, it should play a pivotal role. Trust has been viewed as a basic element in the patient-doctor relationship for a long time. This has been given ethical prominence by Edmund Pellegrino and is embedded in the covenant model of ethical relationships by Tom Veatch; the element of trust is seen as fundamentally significant. If we append Virginia Warren's notion of empowerment, then the distance between doctor and patient shrinks and the relative power dynamics shift.

It may also help to consider Janet Farrell Smith's two principles for ethically adequate communication. These are the principle of equality (including the patient's right to be heard) and the principle of respect (including mutual understanding and empathic communication).[65] With patience and trust laying the foundation, incorporating these two principles should greatly benefit communication around interpersonal violence.

Along these lines, Mary Mahowald observes that a friendship model is helpful. She argues that, "If friendship includes both justice and care, it may serve as an ethical ideal for familial as well as for other interpersonal and professional relationships....Unfortunately, however, care in the health setting is often equated with treatment, and treatment is not necessarily caring."[66] Replacing the concept of "treatment" with that of "caring," may strengthen the level of trust between the violence victim and the physician—and empower the patient in the process.

This is where empathy comes in as well. As expressed by one violence survivor looking back at her recovery from the murderous assault that left her near death:

> Listening is important, listening without judging, or listening without saying you should have something or another or "I know what that feels like," when you really don't. Listening. And if you cannot listen tell them it is too painful to hear, instead of silently turning away. Asking yourself what you can do even if it is not listening. Something simple like sending them a card and telling them you care about them.
>
> A simple act of kindness can validate. Support, knowing that this will take time and your friend may not be herself for a long time.[67]

Victims of violence cannot imagine themselves unscarred. Most wish others had some idea what they've been through and could share in their process of recovery, so they were not alone. We can change all this. We can take steps so they don't feel so isolated or thrown back on their own resources.

We all stand to gain if we do not distance ourselves from those who have suffered acts of violence. More work needs to be done in this area. By getting a fuller grasp of the psychological and metaphysical hurdles they face, we can set in motion changes to the public health system. A much fuller grasp of both the personal and social obstacles facing these patients can help bridge the gap between patient and caregiver. This is a key element in addressing this serious social problem. However, it requires that we shift frames of reference and align ourselves with the patients—the violence survivors—and really see what it is they are facing. As Martha Minow put it in another context: "By aligning ourselves with the "different" person, for example, we could make difference mean something new; we could make all the difference."[68]

Alignment with the victim is necessary if we really want to make systemic change. We can talk about preventative measures and draw up policies intended to transform the horrific spectrum of violence in our society. But such talk is not as likely to succeed if we haven't looked at the healing journey that victims face and the ways in which societal attitudes and prejudices have tainted the healthcare response to violence. A better grasp of the personal and institutional roadblocks that patients must overcome will give us the basis—and the insights—that will then help us address violence prevention.

Nancy Raine writes about a child whose legs were scarred by fire. Evidently the girl, Sarah, was asked in the third grade if she could have one wish, what would it be. Sarah said, "I want everyone to have legs like mine." Reflecting on this wish, Raine observes, "Sarah could not imagine herself without her scars. But she could imagine those scars not setting her apart. She could imagine not being alone. She was not wishing her misfortune on others, but wishing they could share it with her."[69]

We could make all the difference in our response to all of those who have been victimized or traumatized by violence. We could make all the difference in helping those in the response team who suffer secondary traumatization from seeing the toll of violence and putting themselves at risk to help. We could make all the difference by understanding the many obstacles blocking the path to recovery. By looking at the situation from the perspective of the victim—the patient—we see with different eyes. With policies and practices that incorporate that perspective, we can put in place a model of empathy that makes all the difference.

As we have seen, there are many pressing concerns around the treatment of violence survivors by those in public health care, as well as other fields like law enforcement, social services and the like. To develop a more effective and compassionate response on the part of caregivers, we need to make sure we appreciate what the patient is going through. A sense of the complexities of issues that must be examined; consequently, simply dispensing a prescription or handing out a brochure does not provide comprehensive patient care. We have the person's whole life to deal with. A greater appreciation of this fact will help keep a perspective on the importance of shaping a stronger, more humane and caring, response to violence victims.

With such a system in place, we can turn to such matters as the prevention of interpersonal violence, societal attitudes towards perpetrators of violence, the role of the media in desensitization to violence, and so on. That work lies ahead. By taking steps to examine the treatment of violence survivors by health professionals and within the healthcare establishment, we can make a difference. By acknowledging the numerous obstacles they face on both the existential and institutional plane, we can help all those whose lives have been touched by violence.

NOTES

[1] Rebecca Dresser, "What Bioethics Can Learn From the Women's Movement," in Susan M. Wolf, ed., *Feminism and Bioethics, Beyond Reproduction*, (NY: Oxford University Press, 1996), p. 146.

[2] Leigh Turner, "Bioethics, Public Health, and Firearm-Related Violence: Missing Links between Bioethics and Public Health," in Wanda Teays and Laura Purdy, eds., *Bioethics, Justice, and Health Care* (Belmont, CA: Wadsworth Publishing, 2001), pp. 151-52.

[3] Virginia Warren, "From Autonomy to Empowerment: Health Care Ethics from a Feminist Perspective," in Wanda Teays and Laura Purdy, eds., *Bioethics, Justice, and Health Care* (Belmont, CA: Wadsworth Publishing, 2001), p. 52.

[4] Susan Brison, *Aftermath: Violence and the Remaking of a Self*, (Princeton, NJ: Princeton University Press, 2001), p. x.

[5] AMA, *Strategies for the Treatment and Prevention of Sexual Assault,* p. 5, emphasis mine.

[6] Liana B. Winett "Constructing Violence as a Public Health Problem," *Public Health Reports*, 113.6 (Nov 1998). [Journal online]

[7] Fun Gun Facts *http://www.speakeasy.org/wfp/07/Guns9.html*

[8] Sharon M. Wasco, Rebecca Campbell, and Marcia Clark, "A Multiple Case Study of Rape Victim Advocates' Self-Care Routines: The Influence of Organizational Context," *American Journal of Community Psychology*, (Oct 2002). [Journal online]

[9] Barbara Sparks, "Too Scared to Learn: Women, Violence and Education," *Adult Education Quarterly*, 52 (Aug 2002). [Journal online]

[10] June 2001, Bureau of Justice Statistics.

[11] June 2001, Bureau of Justice Statistics.

[12] See Childhood Sexual Abuse: *http://www.wingsfound.org/statsChildhood.html* Year 2000 Findings.

[13] Report to Congress by the Department of Justice, "Stalking and Domestic Violence," NCJ 186157(May, 2001), p. 3·

[14] *Lynne Lamberg*; "Stalking Disrupts Lives, Leaves Emotional Scars: Perpetrators Are Often Mentally Ill," *JAMA*, 286.5 (1 Aug 2001). [Journal online]

[15] U.S. Department of Justice Statistics *http://www.ojp.usdoj.gov.*

[16] Philip J. Cook and Jens Ludwig, "The Costs Of Gun Violence Against Children," *The Future of Children*; Los Altos; Summer 12.2 (2002). [Journal online]

[17] AMA, *Strategies for the Treatment and Prevention of Sexual Assault*, p. 6.

[18] Fiona Bradley, Mary Smith, Jean Long, and Tom O'Dowd, "Reported Frequency Of Domestic Violence: Cross Sectional Survey Of Women Attending General Practice," *BMJ,* 324.7332 (2 Feb 2002). [Journal online]

[19] U.S. Department of Justice, National Crime Victimization Survey, *Criminal Victimization 1999, Changes 1998-99 with Trends 1993-99,* (August 2000), NCJ 182734.

[20] Bureau of Justice Statistics, 2000.

[21] Rosemarie Tong, *Feminist Approaches to Bioethics*, (Boulder, CO: Westview Press, 1997), p. 57.

[22] Susan Brison, *Aftermath: Violence and the Remaking of a Self*, (Princeton, NJ: Princeton University Press, 2001), p. 25.

[23] Elizabeth M. Schneider, "Describing and Changing: Women's Self-Defense Work and the Problem of Expert Testimony on Battering," *Women's Rights Law Report*, 14.2 & 3 (Spring/Fall 1992), p. 236.

[24] See Terry Costlow, "As Stalkers Go Online, New State Laws Try To Catch Up," *Christian Science Monitor*, 3 (Sep 2002), p. 2.

[25] See Robert M. Neil, "Free Speech On The Electronic Frontier," *The Chronicle of Higher Education,* 42.10 (3 Nov 1995), p. A68.

[26] See "E-threats impact workplace," *Security*, 36.5 (5, May 1999).

[27] Lynne Lamberg, "Stalking Disrupts Lives, Leaves Emotional Scars: Perpetrators Are Often Mentally Ill," *JAMA,* 286.5 (1 Aug 2001). [Journal online]

[28] Anonymous, "A Day in The Life of A Stalker's Victim," *BMJ*, 324.7337 (9 Mar 2002). [Journal online]

[29] As quoted by Sandy Banks "Rape Publicity Strikes a Nerve," *Los Angeles Times*, 13 Aug 2002.

[30] Observations of a male victim of gang prison rape, as noted by AMA, *Strategies for the Treatment and Prevention of Sexual Assault*, p. 30.

[31] "Stalking Disrupts Lives, Leaves Emotional Scars: Perpetrators Are Often Mentally Ill," *JAMA*, Chicago; 286.5 (1 Aug 01). [Journal online]

[32] Fiona Bradley, Mary Smith, Jean Long, and Tom O'Dowd, "Reported Frequency Of Domestic Violence: Cross Sectional Survey of Women Attending General Practice," *BMJ*, 324.7332 (2 Feb 2002). [Journal online]

[33] As quoted by Philip J. Cook and Jens Ludwig, "The Costs of Gun Violence Against Children," *The Future of Children*; 12.2 (Summer 2002). [Journal online]

[34] Terry Costlow, "As Stalkers Go Online, New State Laws Try To Catch Up," *The Christian Science Monitor*, 3 (Sep 2002), p. 2.

[35] David Winthrop Hanson, "Battered Women: Society's Obligation to the Abused," *Akron Law Rev.* 27.1 (Summer 1993), p. 33.

[36] Susan Brison, *Aftermath: Violence and the Remaking of a Self*, (Princeton, NJ: Princeton University Press, 2001), p. 21.

[37] Nancy Venable Raine, *After Silence: Rape and My Journey Back* (NY: Three Rivers Press, 1998), p. 228.

[38] See Lynne Lamberg, "Stalking Disrupts Lives, Leaves Emotional Scars: Perpetrators Are Often Mentally Ill, *JAMA*, 286.5 (1 Aug 01). [Journal online]

[39] Susan Brison, *Aftermath: Violence and the Remaking of a Self*, (Princeton, NJ: Princeton University Press, 2001), p. 20.

[40] Quoted by Linda Bell, *Rethinking Ethics in the Midst of Violence*, (Totowa: Rowman and Littlefield, 1993), p. 78.

[41] bell hooks, *killing rage: ending racism.* (NY: Henry Holt and Company, 1996), pp. 51 and 58.

[42] Abby Wilkerson, "Her Body Her Own Worst Enemy," in Stanley French, Wanda Teays, and Laura Purdy, eds, *Violence Against Women: Philosophical Perspectives*, (Ithaca, NY: Cornell University Press, 1998), p. 128.

[43] Cheryl A. Sutherland, Deborah I. Bybee, and Cris M. Sullivan, "Beyond Bruises And Broken Bones: The Joint Effects of Stress And Injuries on Battered Women's Health," *American Journal Of Community Psychology*; 30.5 (Oct 2002). [Journal online]

[44] Quoted by Susan Brison, *Aftermath: Violence and the Remaking of a Self*, (Princeton, NJ: Princeton University Press, 2001), p. 53.

[45] *Ibid.*

[46] Cheryl A. Sutherland, Deborah I. Bybee, and Cris M. Sullivan, "Beyond Bruises And Broken Bones: The Joint Effects of Stress And Injuries on Battered Women's Health," *American Journal of Community Psychology*, 30.5 (Oct 2002). [Journal online]

[47] See Mary B. Mahowald, "On Treatment of Myopia: Feminist Standpoint of Theory and Bioethics," in Susan M. Wolf, ed., *Feminism and Bioethics, Beyond Reproduction*, (NY: Oxford University Press,

1996), p. 106.

[48] AMA, *Strategies for the Treatment of Sexual Assault*, p.23.

[49] AMA, *Strategies for the Treatment of Sexual Assault,* p.26.

[50] Walker and Meloy 1998, p. 142, as quoted by Keith E. Davis, April Ace, and Michelle Andra, "Stalking Perpetrators and Psychological Maltreatment of Partners: Anger-Jealousy, Attachment Insecurity, Need for Control, and Break-Up Context," *http://www.cla.sc.edu/PSYC/faculty/daviske/stalk11davisace.pdf.*

[51] *Ibid*. The research they base this conclusion on is from McFarlane et al. 1999, p. 312.

[52] Michael Dowd, "Dispelling the Myths About the "Battered Woman's Defense: Towards a New Understanding," *Fordham Urban Law Journal,* 19 (1992), p. 582.

[53] Rebecca Dresser, "What Bioethics Can Learn From the Women's Movement," in Susan M.Wolf, ed., *Feminism and Bioethics, Beyond Reproduction*, (NY: Oxford University Press, 1996), p. 145, quoting *New Our Bodies Ourselves*, p. 652.

[54] Sandy Banks, "A Vanishing Point for Rape Victim Anonymity," *Los Angeles Times*, 11 Aug 2002, quoting Gail Abarbanel.

[55] AMA, *Strategies for the Treatment of Sexual Assault*, p.20.

[56] As noted by Elizabeth M. Schneider, "Describing and Changing: Women's Self-Defense Work and the Problem of Expert Testimony on Battering," *Women's Rights Law Report,* 14.2 & 3 (Spring/Fall 1992), p. 217, my emphasis.

[57] Linda Gross, as quoted by Nora Zamichow, "TV Appearance of Teen Victims Suggests Rape Stigma Is Fading Attitudes," *Los Angeles Times*, 8 Aug 2002.

[58] As quoted by Nora Zamichow, "TV Appearance of Teen Victims Suggests Rape Stigma is Fading Attitudes," *Los Angeles Times*, 8 Aug 2002.

[59] AMA, *Strategies for the Treatment of Sexual Assault,* p. 7.

[60] See Fiona Bradley, Mary Smith, Jean Long, and Tom O'Dowd, "Reported Frequency of Domestic Violence: Cross Sectional Survey of Women Attending General Practice," *BMJ,* 324.7332 (2 Feb 02). [Journal online]

[61] Doron J. Kahn, "Attitudes of New York City High School Students Regarding Firearm Violence," *Pediatrics,* 107.5 (May 2001). [Journal online]

[62] Don Manzullo, "Title X Press Conference Remarks," *http://manzullo.house.gov/NR/exeres/A4E36E07-5B14-41AE-BC73-54FCFE0C124D.htm*, 12 Feb 1998.

[63] Martha Minow, *Making all the Difference: Inclusion, Exclusion, and American Law*, (Ithaca, NY: Cornell University Press, 1990), p. 276.

[64] *Ibid*, p. 379.

[65] Janet Farrell Smith, "Communicative Ethics in Medicine: The Physician-Patient Relationship," in Susan M. Wolf, ed., *Feminism and Bioethics, Beyond Reproduction*, (NY: Oxford University Press, 1996), pp. 204-205.

[66] Mary B. Mahowald, "On Treatment of Myopia: Feminist Standpoint of Theory and Bioethics," in Susan M. Wolf, ed., *Feminism and Bioethics, Beyond Reproduction*, (NY: Oxford University Press, 1996), p.

107.

[67] Personal communication to the author.

[68] Martha Minow, *Making all the Difference: Inclusion, Exclusion, and American Law*, (Ithaca, NY: Cornell University Press, 1990), p. 377.

[69] Nancy V. Raine, "Returns of the Day," in Wanda Teays and Laura Purdy, eds, *Bioethics, Justice and Health Care* (Belmont, CA: Wadsworth Publishing, 2001), pp. 62-63.

LAURA PURDY

THE POLITICS OF PREVENTING
PREMATURE DEATH

ABSTRACT: This paper argues that bioethics needs to focus more centrally on political and social practices that contribute to death, disease, or disability. A broader vision would help bioethicists show why
attacking the root causes of premature death and physical suffering should have priority over contemplating the pros and cons of expensive, risky, high-tech solutions that often address problems after they have
developed. Only this approach is consistent with bioethicists' professed values, such as the wrongness of
hastening patients' deaths without their consent. This approach puts corporate and governmental decision-
making at the center of moral inquiry, as they determine economic, environmental, and occupational conditions that are a major source of death, disease, and disability. However, there are significant cultural and
institutional obstacles to this kind of bioethics.

KEY WORDS: Death, premature death, physical suffering and ethics, informed consent, moral decision-
making in bioethics.

Most people value long and healthy life. Bioethics attempts to think critically about
the moral issues connected with this goal.

Although bioethics has existed for as long as people have been contemplating
these issues, it did not become an academic "field" until some thirty years ago.
During this time it has expanded tremendously, and it has now begun to generate critiques of its own topics and methods.

One such critique holds that the now "standard" issues, the ones that show up in
the most widely-used texts, have arisen from relatively narrow concerns, and are
based on unexamined moral and political world-views[1]. Thus, for example, gallons of
ink are expended on what are known as "end-of-life issues," and the exciting new
technologies connected with genetics and reproduction, while hugely worrisome phenomena, such as the growing numbers of uninsured individuals in the U.S., remain
relatively unnoticed.[2] Nor has mainstream bioethics focused as much on the differential treatment of white women and members of disadvantaged minorities as one might
expect, despite mounting evidence of its sometimes devastating consequences for
their health.[3]

This paper builds on these critiques, and argues that bioethics has failed its mis-

M. Boylan (ed.), Public Health Policy and Ethics, 167-185.
© *2004 Kluwer Academic Publishers. Printed in the Netherlands.*

sion in yet another serious way: it focuses far too much on the medical environment at the expense of non-medical practices that contribute to death, disease, or injury. In short, much more work needs to be done on the ethics of public health.[4]

THE PROBLEM

Like most bioethics scholars, I have been drawn to relatively narrow questions, despite increasing unease about the overall state of the world. Not wanting to be pried loose from the issues that gripped me, it seemed to me that although much was seriously wrong, there wasn't anything I could do about it, qua bioethicist. Yes, people, children and women particularly, are getting sick and dying, primarily because of conditions generated by poverty. My view was that countries where this is happening should allocate resources to stop the deaths, and rich nations should help, too. End of story.

Only recently did I realize what more might be said. The crystallizing moment came when I confronted researchers who wanted to pursue a difficult, expensive, and potentially risky technology, "to save lives."[5] The technology in question, xenotransplantation, proposes to use animal tissue and organs to cure chronic disease and end-stage organ failure. However, the technical obstacles are formidable. "Ordinary" human-to-human transplants are sufficiently problematic because of the need for powerful drugs to control the body's rejection of "foreign" tissue. Animal-to-human transplants involve such violent rejection mechanisms that creating and cloning transgenic animals appears to be necessary to foil them. Worse still, there is an unknown but potentially significant risk of transmitting disease from the sacrificed animal to the human host, and thence on to the public. The worst-case-scenario is the creation of an unstoppable, new, virulent, lethal disease.

The extraordinary resources needed for xenotransplantation research, together with the potentially significant risk to public health, seemed to me excellent reasons for seeking alternative ways of attempting to save the lives of patients with end-stage organ failure.[6] Reframing the "organ shortage" in terms of a gap between the supply of organs and the number of those who could benefit from them creates new ways of conceptualizing the problem, such as focusing on prevention, developing creative new strategies to increase human donation, alternative treatments, and new technologies (such as stem cells).[7]

This position was met with considerable negativism. I was just a theoretician, far removed from the clinical reality of dying patients, alternatives were pie-in-the-sky, my approach was "unscientific," progress always involves risk, etc., etc. etc. At the end of a grand rounds at which I argued my case before a group of research-oriented health care providers, the moderator asked the audience whether they would want xenotransplantation available to treat a loved one. More than two-thirds of the audience raised their hands.

When I suggested that not every technology needs to be pursued (not a new argument), and that some risks constitute grounds for a moratorium on research (common sense), the trump card continued to be saving lives. But that suddenly seemed a myopic response, given that worldwide, many lives could be saved for pennies apiece, and at no risk. Why was it rational to try to save the lives of those with end-stage organ failure, seemingly no matter what the potential cost (even quite possibly at the cost of other lives down the road), when there are cheap, safe, and proven remedies for the global epidemic of premature deaths?

The most obvious response is that transplant surgeons (and other xeno researchers) are not in development, and so cannot prevent *those* deaths. But that answer raises the question whether each profession should prevent deaths in its own way or whether society needs an overall plan that sets priorities for this activity. I believe that only such an approach could ensure that inefficient or potentially risky programs are not undertaken.

However, in one important response, a leading scholar conceded that the push for xeno is "industry-driven."[8] Indeed, in 1998 it was suggested that there could be a $6 billion market in xenotransplant by 2010.[9]

Although I had been becoming increasingly aware of the influence of commercialization in science and medicine, this answer led me to see the question at hand in a whole new way. Although I've always been a bit puzzled by the intense focus on end-of-life issues, and had thought that I had taken to heart Steven Miles's admonition about the priorities of bioethics—yet another, still more fundamental dimension of the issue opened up before me.

On the one hand, the end-of-life discussion (and most law) is predicated on the assumption that it can be wrong to hasten death by an hourBeven, some believe, when requested by the individual in question. On the other, bioethicists[10] have, for the most part, been silent about the fact that diarrhea, malaria, measles, pneumonia. HIV/AIDS and malnutrition—all preventable—kill almost eleven million babies and children every year.[11] And that 515,000 women die every year during pregnancy or childbirth when simple preventive measures could save most of them.[12] Talk about involuntary, hastened deaths.[13]

So why isn't this as big a topic as deathbed decision-making? Is it that, as I initially surmised, there are no moral knots to disentangle, no work here for bioethicists?

Or could it be instead that the morally relevant differences between these cases is so apparent that seeing them as equivalent is evidence of moral incompetence?

Exploring, for a moment, the second supposition, what glaring differences might there be? Perhaps we owe less to citizens of other countries. But if saving lives is truly a top priority, why does nationality matter?[14] Do bioethicists really think that the values guiding the end-of-life debate do not apply to aliens living in the United States? Or, could it be that although it is—other things being equal—wrong to deliberately kill foreigners, that there is no duty to prevent premature deaths? It is hard to believe that this slender theoretical reed could support such a moral burden. In addition, to the extent that the United States (and the Western world, generally) have policies that promote or sustain impoverishment or violence in the Third World, it is disingenuous to act as if such deaths are just acts of "nature."

Underlying some such "respectable" arguments are (no doubt) a variety of less savory premises. Racism, for instance, discounts the deaths of people of color. And sexism devalues women's lives.

In the absence of any other plausible explanation for current attitudes, further exploration is in order. To avoid the moral issues raised by nationality, let us now turn to the United States.

PREVENTING PREMATURE DEATH IN THE UNITED STATES

A study published in the *Archives of Internal Medicine* found that 60,300 workers were killed by occupational illnesses, and 862,200 workers became ill from them in 1992. In addition, 6,500 workers were killed in work-related accidents, and 13.2 million were injured that year. Occupational injuries and diseases cost at least $171 billion that year.[15]

In addition, it is estimated that there are at least 28,000 deaths, and 130,000 serious injuries every year because of dangerous consumer products.[16] These yearly figures do not include the pervasive environmental pollution to which most of us are now exposed. A 1998 study conducted at Cornell University by David Pimentel concludes that "an estimated 40 percent of world deaths can now be attributed to various environmental factors, especially organic and chemical pollutants."[17] Pimentel argues that disease is promoted by malnutrition and environmental pollutants, and that global climate change will reinforce their effects. For example, some 6,000 annual deaths from lung cancer in the U.S. may be caused in North America by power plant emissions (tiny soot particles, sulphur dioxide and nitrogen oxides).[18]

The scope of such exposures is enormous. According to Peter Montague, the chemical industry doubles in size every twenty years. In 1995, 70,000 chemicals were in the marketplace. Seventy percent of them had not been tested at all for harm to human health, and only two percent had been fully tested. In addition, over a thousand new, mostly untested, chemicals are introduced every year.[19]

The media mostly ignore such statistics. As John Judis points out, "this issue is the all-time unsexy news story. It only gets attention when a mine explodes or a postal employee goes berserk."[20] Ongoing loss of life is less newsworthy than the sports coverage that fills up so much of the average newspaper.[21] This superficial, crisis-oriented approach tends to support the comfortable belief that these deaths and injuries are, like those in the Third World—"natural" and inevitable, just the downside of progress, just the cost of doing business in the contemporary world, creating an environment that dampens awareness, curiosity, or concern on the part of the average citizen.

What the media fail to report is that government regulation of workplaces can make a big difference in these numbers. Judis notes:

> During the 1970s and the late 1990s, when the Occupational Safety and Health Administration (OSHA) was adequately funded, it had a dramatic effect on the incidence of job injuries. Between 1995 and 2000, for instance, the number of injuries and illnesses dropped to 6.1 from 8.1 for every 100 workers (Judis, 10).

Broader environmental and safety regulation can also make a big difference: it got the lead out of gasoline, and made cars much safer.[22]

In short, corporate decision-making (and decisions by governments about how to regulate those decisions) have long determined whether people (workers, consumers, and the public) live or die.[23] For example, businesses have knowingly exposed workers (and local residents) to toxic substances like silica, and asbestos.[24] Ford manufactured the Pinto, knowing that people who drove it could be burned alive if the car were hit from behind, even though the car could have been made safe by the addition of an $11 part.[25] Automakers pushed for the addition of lead to gasoline to increase the power of engines, even though it had been known since ancient times that lead is a deadly toxin.[26]

Similar questionable decisions are being made today. For example, bovine growth hormone increases the levels of IGF-I (insulin-like growth factor-1), a growth-promoting substance that is the same in cows and humans, in dairy products.[27] Yet, studies have shown that premenopausal women with the highest levels of IGF-I in their blood are seven times more likely to contract breast cancer than those with the lowest levels, and men with the highest levels of the hormone in their blood are four times more likely to get prostate cancer.[28] This decision is at best questionable. Likewise, the government appears to be on the verge of allowing radioactive waste to be recycled into consumer products, even though radiation is a known carcinogen.[29]

Are corporations morally or legally responsible for such decisions? One might initially distinguish at least four different cases. First, a business might do thorough research before going into production, discover that there is already much known about it, that there is neither evidence of significant risk nor any theoretical reason to fear harm. Second, a business might either fail to do the relevant research, or discover that little is known about it, and go forward in ignorance. Third, a business might do the relevant research, discover reasons for thinking that its proposed course of action could be harmful, but choose to go forward because of expected profit. Last, and perhaps most often, (1) or (2) is the case, but as the years pass, evidence of harm becomes available.[30]

It seems clear that there is no moral culpability in the first case, although justice requires eradicating the newly-recognized danger, as well as restitution for those harmed. The other cases are a different story. As the 19th century writer William Clifford argued so eloquently, you cannot evade responsibility for harm to others just because you did not inform yourself about the risks. If you do discover that your proposed course of action involves risk to others, you cannot escape responsibility for ensuing harm.[31]

Although Clifford is a plausible starting point for thinking about corporate responsibility, most courses of action entail some risk, and his work needs to be refined to consider when risk is acceptable and to determine degrees of responsibility. But if a corporation exposes its workers to a known carcinogen to avoid the need for expensive changes in production practices, it is reasonable to ask whether that is morally different from killing an elderly relative for an inheritance. One might also reasonably ask why corporate homicide doesn't focus attention on the concept and theory of the corporation, just the way elder-murder focuses it on proposals to legalize euthanasia.

THE WORLD OF CORPORATIONS

The answer must be that there are, after all, morally relevant differences between these acts. One proposal for such a difference is that the firm's activity benefits the community by providing jobs and useful products, and so cost-benefit analysis will always favor its decisions. However, this position must be subjected to critical scrutiny, given the long history of such claims that turned out to be false.[32] Other common candidates would be that "corporate persons," unlike real ones, cannot be responsible for their acts, that risking harm is morally different from deliberate killing, and that it is wrong to target specific persons, but not vulnerable populations.

There is some literature on these interesting questions,[33] but starting the discussion by focusing on them deflects us away from what I see as the basic questions. What are corporations, anyway, and why can they get away with what looks like killing, even serial killing?[34] And why is the default assumption that—of course—the alleged morally relevant differences between their acts and "ordinary" murder aren't worth serious attention, and thus the commonsense conclusion must be that they are not murdering people? And, given the consequences of their acts, why isn't the burden of proof on showing why what they do is not murder, rather than on showing why it is murder?

My own investigation into these questions has been eye-opening. Corporations started as highly-regulated associations created for specific, limited purposes. Ensuing theoretical and legal decisions have now made them free-wheeling legal persons, with many of the constitutional rights formerly reserved for human persons. For example, corporations now enjoy First Amendment rights that protect both commercial and political speech (where "speech" includes a right to make political donations).[35] Oddly, there is deep legal skepticism about whether they could have legal responsibilities. Yet even the most conservative political philosophy recognizes the inherent ties between rights and responsibilities, and that having rights without responsibilities is a recipe for trouble.

How could this obviously problematic state of affairs have come about? One strand of the story appears to arise from the University of Chicago theorists Milton Friedman, Frank Easterbrook, and Daniel Fischel. Friedman argues that although businesses must respect the law, seeking profits is their sole responsibility. Easterbrook and Fischel extend this line of reasoning, contending that corporations should break the law whenever it is profitable to do so. Any resulting fines are just the cost of doing business.[36]

Conceptualizing regulation as a mere obstacle to profits generates the obvious next step: trying to influence the law to reflect perceived corporate interests.[37] Industry groups now invest large sums in lobbying legislators who can affect whether regulations are passed or enforced.[38] They also create apparently objective and neutral non-profit think-tanks and organizations to articulate and promote their views.[39] The revolving-door between industry and regulatory bodies is well known.

Equally important, the whole issue is seen through the lens of class interest that evaluates crimes, at least in part, on the basis of who commits acts rather than their consequences. As sociologist Edwin Sutherland noted in 1939, "crime" is defined as street crime—direct, violent, harmful assault, the kind of assault that tends to be committed by poor, uneducated persons. Harmful acts committed by respected middle- and upper-class individuals are "white-collar" crime, whose seriousness is discounted.[40] In fact, it was not until the nineteenth century that some harmful acts committed by businesses were even criminalized. Yet, despite the fact that such crimes kill far more people than individual criminal activities,[41] they are still barely seen as crime, and generally are punished leniently (Vago, p. 242).[42] In short, there is class privilege in both the definition of crime and the way it is punished.[43]

Further inquiry along these lines is obviously necessary to help society understand harmful corporate behavior. It takes us far astray from bioethics' usual haunts, into political philosophy, history, especially legal history and theory, political science, economics, and sociology. Because of the way scientific studies are used in both corporate decisions and regulatory agencies, bioethics must also take on the minefields of so-called "junk science,"[44] and theories about measuring and evaluating risk.[45] All this may seem daunting, but individual bioethicists have already accomplished comparable tasks. In any event, only such a comprehensive approach could show that reducing the toll of death, disease, and injury from corporate activity and the failure to regulate it is required by principles and values already widely accepted in other contexts.

Unfortunately, there are major obstacles to this proposal. One is the charge that such inquiry is "political," and thus unsuitable for scholarly pursuit. Risking this judgment at a time where "politicization" is so feared that one's standing in the field could plummet, possibly costing one a coveted job, promotion, or grant.[46]

I believe that the charge that this kind of inquiry is inappropriately political is groundless. In fact, it seems clear that the bioethics status quo is itself "political." The field is rooted in moral and political assumptions and institutions that are—despite increased recent self-scrutiny—still largely taken for granted.[47] A consistent and humane bioethics must face this fact.

"POLITICIZATION" AND OTHER OBSTACLES
TO A CONSISTENT AND HUMANE BIOETHICS

Although charges of "politicization" can be very successful at blocking particular paths of inquiry, it is striking how poorly defended they have generally turned out to be.[48]

People rarely define the central terms "political" and "politicization." It is implied that "political" discourse relies on manipulation or sleazy reasoning instead of theory, evidence, and logic. But that accusation could be valid, if at all, only against a particular way of conducting a discussion, not a particular line of inquiry.

"Political" discourse is often accused of relativism, but that charge can hardly apply to well-argued moral claims. Sometimes charges of "politicization" appear to imply that there is some distinction between the "political" and the "moral," such that some questions (the nature of the good life?) are within the scope of scholarship, but that there is no unbiased way of investigating basic social institutions (upon which, in reality, the possibility actualizing the good life rest). But then it would appear that Plato and Aristotle (not to mention dozens of other theorists) were on the wrong track, too. Any such accusation boils down to a demand that basic human institutions are beyond criticism. Although this position may further the interests of those who benefit from their current configuration, it is a transparent attempt to preserve privilege by deflecting scholarly attention from that fact.

It follows that the sort of work I am pressing forBfollowing the trail of mortality and morbidity to its origins—is not political in any of these malignant senses. It aims at a coherent position, seeks evidence, rejects fallacious reasoning, and takes objections seriously. It proposes to examine the corporation, a crucial player in the contemporary world, trace its theoretical underpinnings, consider how it is constructed, the consequences of that construction, and possible alternatives. It is critical of the status quo, but only an already inappropriately politicized area of scholarship could reject it on those grounds.

This work is, of course, a normative enterprise: it is based on particular conceptions of justice and human welfare. But so is bioethics-as-usual. The problem is that the underpinnings of bioethics-as-usual are so taken for granted that they are seen as neutral, their normative nature unnoticed.[49] Such unevaluated foundations affect our work in subtle but crucial ways, determining where we see burdens of proof, perhaps inclining us to accept weaker arguments for favored theses, and demand stronger arguments for disfavored ones.

Moreover, understanding how social institutions influence thought has been steadily growing, especially in science. Yet bioethics is still relatively untouched by this new knowledge; the field must focus more consistently on it to fulfill its promise.

Of course, there is now a tradition of worrying that bioethics is floundering, detached from its moral moorings.[50] One recent critic, Helen B. Holmes, maintains that there is a bias toward overly universal and abstract work, overvaluing individualism and autonomy, lack of awareness of broader social movements, the tendency toward uncritical acceptance of assumptions, a crisis orientation, and false dualisms.[51]

The charge of inappropriate individualism is a mainstay of the progressive critique of liberalism. Classical liberalism holds individuals responsible for their choices, ignoring the underlying constraints on those choices (and *their* consequences) created by social and political arrangements. For example, workers who take jobs in obviously dangerous situations are judged to have voluntarily accepted any risk associated with them, despite the fact that they may have had no alternative. It also ignores the fact that more stringent regulation of workplaces could in many cases substantially lower the risks to which workers are exposed. So the progressive critique melts away the alleged morally relevant difference between the two types of premature death we have been considering ("end-of-life" deaths and those hastened by preventable exposure to dangerous conditions). It should follow that the latter deserve at least as much attention that the former.

This issue is closely related to Holmes's criticism that bioethics tends to accept prevailing social and political assumptions. Scholars like to think they are immune from intellectual fashion, but history shows how mistaken this belief can be. It may be especially difficult to distinguish intellectual fashion from truth when supporting corporate or governmental policies can lead to publications, substantial financial rewards, and powerful, prominent positions, but criticizing them risks one becoming the target of lawsuits, smear campaigns, or attempts to suppress one's work.

This issue is connected both with the way bioethics is developing as a discipline, and by the increasing commodification alluded to early on in this paper.

One key element is the rise of clinical bioethics. Clinical bioethics emphasizes the importance of physicians, and clinicians, generally. Unsurprisingly, it also tends to take the parameters of the health care environment for granted, and attempts to resolve problems within that context.

Clinicians—and physicians, in particular—have always played an important role in bioethics, and rightly so. Bioethics is inherently interdisciplinary, and health care practitioners should be central players in medicine; their participation is needed to help outsiders understand their perspectives, to point out problems that outsiders would otherwise be unaware of, and to provide technical information. When clinicians become bioethicists, however, they tend not to undertake the kind of full humanities education represented by the Ph.D., but rather shorter courses that are necessarily less comprehensive. Where physician-bioethicists are in charge of the bioethics enterprise, therefore, the focus will most likely be on relatively immediate and practical issues, to the detriment of more basic assumptions. The clinical environment rewards this approach, even where bioethicists have traditional humanities backgrounds. Studying these practical matters is immensely important but not the whole story.

Two other factors limit the scope of inquiry. One is that even individuals with extensive humanities education may not have the background needed to take on contemporary issues. Philosophy programs, for example, might still focus on traditional moral and political philosophy. Although many of the issues they cover are still relevant (individual vs. state, the nature of political authority) they do not help students understand contemporary developments, such as the rise of the multinational corporation or trade issues. Yet it is arguable that the rise of corporate wealth and power is at the heart of some of the most pressing problems in bioethics. The other factor is that graduate programs may encourage students to stay within well-trodden intellectual paths instead of questioning the basis for the curriculum. After all, graduate programs are generally more geared to preparing young professors than helping students ponder their moral and political premises.

A second major problem connected with the rise of clinical bioethics is its setting in the clinic. The values and assumptions of the clinic may be quite different from those of university humanities departments. For example, clinician-researchers do not necessarily have tenure and thus lack the kind of job security that encourages people to resist inappropriate political pressure. They are more dependent on the evaluation of their peers to maintain their positions, get the grants they need to keep up their research programs, or publish. Although such peer review can ensure quality, it can also be a vehicle for squelching scholarly or political nonconformity.

Because of this vulnerability, other more general features of the academic environment may be especially limiting in clinical settings, although researchers in traditional university departments (especially untenured ones) are not immune from them. Where publication is the chief measure of productivity, quantity is easier to evaluate than quality. Never have there been more opportunities to publish, or more pressure to do so. This state of affairs discourages people from taking on big, open-ended projects, leading them instead to pick more circumscribed ones that they know they can publish quickly. This tendency is especially regrettable now. One reason is that contemporary problems cry out for bold, but intellectually risky, interdisciplinary approaches. The other reason is that it has never been easier to undertake this kind of work, given the availability of electronic databases and full-text electronic sources that make it possible to cover ground in months that would previously have required years. At the same time, this pressure to publish, in conjunction with the burgeoning literatures in many fields, means that it is becoming increasingly difficult to ensure that one has taken account of all relevant work before publication.

Other pressures are at work, too. The push for usable outcomes, whether new technologies or ethical solutions to common problems, is a two-edged sword. On the one hand, they can enhance human welfare immediately. On the other, more fundamental research that functions as a resource for the future can get crowded out.

This danger is exacerbated by the new economics of academe. The contemporary legal infrastructure that opened up possibilities for patent protection and commercial development of university research products has radically changed the scene. As David Shenk writes in his thought-provoking treatment of the subject:

> The infusion of private capital is staggering. In 1997 US companies spent an extraordinary $1.7 billion on university-based science and engineering research, a fivefold increase from 1977. More than 90 percent of life-science companies now have some type of formal relationship with academic scientists, and 60 percent of those report that they have achieved new patents, products and sales as a result. In the realm of university science, at least, that once-remote ivory tower now finds itself cater-corner to an office park—in many cases literally.[52]

The hope is that wonderful new technologies will flourish as a result of these developments, but the costs appear to be significant. A creeping culture of protectionism and secrecy seems to be replacing the traditionally open culture of science, raising fears about the future of scientific progress. The consequences for bioethics may be equally problematic. At the individual level, conflicts of interest are now pervasive for scientists, potentially affecting the quality of information bioethicists must work with. But bioethicists may be even more directly affected by conflicts of interest arising from their own funding sources. For example, if their workplace has a financial interest in the success of a new technology, they may feel uncomfortable pointing out serious problems with it. At the broader institutional level, increasing corporate funding already appears to be jeopardizing universities' support for "counterproductive" scholarship (such as criticism of genetically modified organisms) or commercially worthless scholarship (most research in the humanities, and work in other areas focusing on beneficial projects with little potential for profit).[53]

CONCLUSION

Greater understanding of the context of bioethics should help many bioethicists engage in the broader, morally and politically aware scholarship I argue for here. It should help them more clearly flag (for later examination) the assumptions they rely on when they study practical problems. It should also encourage them to consider public health dimensions of such problems. Thus, for example, scholars who consider the ethics of research on new technologies, like xenotransplantation, need always to consider what role their development might play in human welfare, worldwide. They also need to be sensitive to how their personal interests might be influencing their conclusions.

Also, of course, I hope that my arguments will encourage bioethicists to take seriously urgent, politically-charged issues in public health ethics like premature death and disability. It might seem that my quest here implies an imperialism that subordinates all other values to preventing premature death, disease, or injury. That is not the case. My dream is for a more comprehensive bioethics, one that dares apply its principles and values beyond the relatively narrow contexts to which it has been too often confined. It is rooted in the conviction that unnoticed ethical and political assumptions and influences blind us to the obvious point, that life and health are rightly at the top of most people's priorities. These forces dispose us to accept that the burden of proof should be on those who doubt that the premature deaths discussed here are "necessary" or are far less culpable than those that could be caused by wrong end-of-life decisions. My thesis is not that every attempt to re-categorize deaths that are now taken to be morally dissimilar will succeed, but rather that the ethical and political assumptions embedded in our culture are at work here in problematic ways. Only determined excavation can help us notice them and, where necessary, reframe the questions we ask.

The key issue is that we must let go of the comfortable assumption that most disease and death is either inevitable (a result of disease processes or genetic programming that we do not yet understand or control), or a result of autonomous choices (to smoke, to be sedentary), and focus squarely on the social determinants of health.[54] Instead, we must be committed to learning how much of our lives that we now accept as "natural" and "inevitable" is instead constructed by our own decisions and those of others, including traditions and institutions.[55] I have argued that this outlook forces us to cast a new look at corporate behavior and governmental failure to regulate it to protect life and health.[56] But the inquiry must go still deeper, requiring each of us to take responsibility for the trade-offs implicit in all our choices, and to notice—and act on—constraints that cause us to choose unwisely. Everything must be in question: what we eat,[57] how we shelter ourselves, how we travel, which inequalities we tolerate and at what cost.[58] The outcomes affect not only our own health, and that of our fellow citizens, but also the health of humans all over the world [59]

Only a bioethics that pushes us toward this inquiry can be consistent and humane. And such inquiry is required by the public interest. Therefore, institutional environments must be altered to encourage such discomforting inquiry to flourish.

NOTES

[1] For social science perspectives on bioethics, see Arthur Kleinman, "Anthropology of Bioethics," *Writing at the Margin: Discourse Between Anthropology and Medicine* (Berkeley, CA: University of California Press, 1995); *Social Science Perspectives on Medical Ethics*, ed. George Weisz (Philadelphia: University of Pennsylvania Press, 1990); and Raymond DeVries and Janardan Subedi, eds., *Bioethics and Society: Constructing the Ethical Enterprise* (Upper Saddle River, NJ: Prentice Hall, 1998).

[2] See Steven Miles, "The Role of Bioethics and Access to US Health Care: Is Bioethics One of Kitty Genovese's Neighbors?" *Bioethics Examiner* 1.2 (Summer, 1997): 1-2.

[3] See, for example, Susan Sherwin, *No Longer Patient: Feminist Ethics and Health Care* (Philadelphia: Temple University Press, 1992).

[4] For a helpful summary of the issues, see Lawrence O. Gostin, "Law as a Tool to Advance the Community's Health," Part 1 of Public Health Law in a New Century, *JAMA*, 283.21 (June 7, 2000): 2837-41. For recent bioethics work in public health, see Dan E. Beauchamp and Bonnie Steinbock, Eds., *New Ethics for the Public's Health* (NY: Oxford University Press, 1999).

[5] "Preventing death" seems generally preferable to "saving lives" because the latter expression seems to leave more questions. What conditions must be met to justify the claim that a life has been saved? How long does the individual in question need to survive? What about quality of life? For an exploration of the concept of life expectancy, see Robin Small, "The Ethics of Life Expectancy," *Bioethics* 16.4 (2002): 307-334.

[6] My focus is on whole organs, rather than cells or tissues that might be used to treat chronic diseases like diabetes or Alzheimer's. All forms of xenotransplantation raise additional moral issues not touched on here, such as animal welfare, informed consent, and access.

[7] See my "Should We Put the 'Xeno' in 'Transplant'?" forthcoming in *Politics and the Life Sciences*.

[8] A.S. Daar, e-mail, May 5, 2000.

[9] Peter Laing, cited by Declan Butler, Briefing: "Xenotransplantation," *Nature* 391 (January):325.

[10] For work in this vein, see *World Hunger and Moral Obligation*, ed. William Aiken and Hugh LaFollette (Englewood Cliffs, NJ: Prentice-Hall, 1977). See also the recent *Living High and Letting Die: Our Illusion of Innocence*, by Peter Unger (New York: Oxford University Press, 1996), and Rebecca J. Cook and Bernard M. Dickens, "The Injustice of Unsafe Motherhood," *Developing World Bioethics* 7.2 (2002): 64-81.

[11] Cited in Elizabeth Olson, "U.N. Says Millions of Children, Caught in Poverty, Die Needlessly," *New York Times*, March 14, 2002 (www.nytimes.com).

[12] "One Woman dies in Childbirth Every Minute, UNICEF Notes," kaisernetwork.org Daily Reports, *http://www.kaisernetwork.org/dailyreports/rep index.cfm?DR ID'9934*.

[13] "Hastened" or "premature" death is in need of further definition. But the cases at issue here are very clear.

[14] See Satinah Sarangi, "Corporate Violence in Bhopal." Paper read at the International Conference on Preventing Violence, Caring for Survivors, Role of Health Profession, and Service in Violence, organ-

ized by the Centre for Enquiry into Health and Allied Themes (CEHAT), November 28-30, Mumbai, India. If Sarangi's account of the consequences and probable causes of the Bhopal disaster is accurate, it is hard to take seriously the claim that the nationality of the victims means that their deaths are less morally important. Accessed at *http://www.altindia.net/bhopal/Corpviolence.htm* on August 15, 2002.

[15] J.P. Leigh et al., "Occupational Injury and Illness in the United States. Estimates of Costs, Morbidity and Mortality," *Archives of Internal Medicine* 157.14 (July 28, 1997):1557-68, cited in "1997 Snapshots, Part 1," *Rachel's Environment & Health Weekly #578*, December 25, 1997. *http://www.monitor.net/rachel/*. Further statistics can be found on the "Injuries, Illness, and Fatalities Homepage," U.S. Bureau of Labor Statistics, *http://bls.gov/IIF.* For an extremely comprehensive report, see *Death on the Job: The Toll of Neglect*, 11th edition, April 2002, provided by the AFL-CIO Safety and Health Department, available on the web at *http://www.aflcio.org/safety/deathonthejob.pdf,* accessed on August 15, 2002.

[16] Donald J. Miester, Jr., "Criminal Liability for Corporations that Kill," *Tulane Law Review*, March 1990, vol. 64:919. Lexis-Nexis Academic Universe.

[17] David Pimentel, "Ecology of Increasing Disease: Population Growth and Environmental Degradation," *BioScience*, October 1998; http://dieoff.com/page165.htm, accessed August 15, 2002.

[18] For every 10-(mu)g/ml elevation in fine particulate air pollution, there is about a four percent increased risk of death from all causes, a six percent rise in cardiopulmonary deaths, and about eight percent rise in deaths from lung cancer. C Arden Pope III, Richard T. Burnett, Michael J. Thun, Eugenia E. Calle et al., "Lung Cancer, Cardiopulmonary Mortality, and Long-Term Exposure to Fine Particulate Air Pollution," *JAMA* 287.9 (March 6, 2002):1132-41.

[19] Peter Montague, "Environmental Trends," *Rachel's Environment & Health Weekly #613*, August 27, 1998; available at http://www.rachel.org/bulletin/index.cfm?St'4; accessed August 15, 2002.

[20] John B. Judis, "Hidden Injuries of Class: Undermining OSHA by a Thousand Cuts," *The American Prospect*, (June 3, 2002): 10.

[21] The problem with the media is not just the failure to provide coverage of important ongoing issues. Still more worrisome is its willingness to present citizens with items prepared by third parties as objective reporting, often with little or no notice. See Sheldon Rampton and John Stauber, *Trust Us, We're Experts!* (New York: Jeremy P. Tarcher, 2001), chapter one. They cite studies showing that 40-50 percent of items in mainstream media can be constituted by such sources (pp. 22-24).

[22] Stephen L. Isaacs and Steven A. Schroeder, "Where the Public Good Prevailed: Lessons from Success Stories in Health," *The American Prospect*, (July 4, 2002): 26-30.

[23] For a good introduction to the moral issues connected with corporations, see Peter A. French, *Corporate Ethics* (Orlando: Harcourt Brace and Company, 1995). For a more basic introduction, see Mathew H. Shaw and Vincent Barry, *Moral Issues in Business*, 8th edition (Belmont, CA: Wadsworth, 2001).

[24] See Rampton and Stauber, chapter 4; for an example of the meat-packing industry's treatment of workers, see Eric Schlosser, "The Chain Never Stops," *Mother Jones*, (July/August 2001): 38-47, 86-87.

[25] French, *Corporate Ethics*, p. 129. Ford's reasoning was apparently that it would manufacture 12,500,000 cars. 2,100 might catch fire, leading to 180 deaths and a further 180 serious burn cases. Because Ford valued each of the lost lives at $200,000, its cost-benefit analysis concluded that it should pay the damages rather make the care safer.

[26] Rampton and Stauber, p. 90. Leaded gasoline released about 30 million tons of lead into the environment in the U.S. Rampton and Stauber, p. 93.

[27] See Shiv Chopra et al., "RBST (Nutrilac) 'Gaps Analysis' Report by RBST Internal Review Team, Health Protection Branch, Health Canada," Ottawa: Health Canada, April 21, 1998. Health Canada is the Canadian equivalent of the U.S. Food and Drug Administration. The report can be accessed at www.nfu.ca/gapsreport.html. See also Peter Montague, "Milk Controversy Spills into Canada," *Rachel's Environment & Health Weekly*, #621, October 22,1998; http://www.rachel.org/bulletin/index.cfm?St'4; accessed August 15, 2002.

[28] See Susan E. Hankinson et al., "Circulating Concentrations of Insulin-Like Growth Factor 1 and Risk of Breast Cancer," *Lancet*, 351(9113) (May 9, 1998): 1373-75, and June M. Chan et al., "Plasma Insulin-Like Growth Factor-1 and Prostate Cancer Risk: A Prospective Study," *Science* 279 (January 2, 1998): 563-66.

[29] Susan Q. Stranahan, "Radioactive Recycling," *Mother Jones*, (July/August 2002): 15-16.

[30] Corporations may see attempts to regulate as a threat to immediate profits, and use obstructive measures to delay or prevent implementation, even when the evidence for harm starts to mount. Thus, for example, the automobile and power industries have brought suit against proposed EPA regulations that would reduce the pollutants that claim 6,000 lives a year. (See Andrew C. Revkin, "Soot Particles Strongly Tied to Lung Cancer, Study Finds," *New York Times*, March 6, 2002 (internet edition)). They also fund individuals and organizations to produce studies that support their own positions. (See Rampton and Stauber, chapter 8.)

[31] William K. Clifford, "The Ethics of Belief," *The Ethics of Belief and Other Essays* (Buffalo, NY: Prometheus Press, 1999); O.P. 1877.

[32] Rampton and Stauber recount the story of vinyl chloride: "Faced with proof that vinyl chloride caused a rare form of liver cancer, chemical manufacturers announced that a proposed federal standard for vinyl chloride exposure would cost two million jobs and $65 billion. . . . .After the screaming was over, the standard was adopted and the industry continued to flourish, without job losses and at 5 percent of the industry's estimated cost (p. 87). Additional examples abound. The automobile industry spent twenty years and more than twenty million dollars fighting the introduction of airbags. (See Steve Nadis and James J. MacKenzie, *Car Trouble* (Boston: Beacon Press, 1993), p. 54.

[33] My own exploratory work was in "Abortion and the Argument from Convenience," *Reproducing Persons: Issues in Feminist Bioethics* (Ithaca, NY: Cornell University Press, 1996), pp. 132-45. See also Alastair Norcross, "Comparing Harms: Headaches and Human Lives," *Philosophy and Public Affairs*, 26.2 (1997): 135-167, especially p. 160ff.; Michael Ridge, "How to Avoid Being Driven to Consequentialism: A Comment on Norcross," *Philosophy and Public Affairs*, 27.2 (1998): 50-58; and Alastair Norcross, "Speed Limits, Human Lives, and Convenience: A Reply to Ridge," *Philosophy and Public Affairs*, 27.2(1998):59-64.

[314] A key issue is raised by Richard Grossman, "Can Corporations be Accountable? Part 2," *Rachel's Environment & Health Weekly #610*, August 6, 1998; http://www.rachel.org/bulletin/index.cfm?St'4; accessed August 15, 2002.. Grossman points out that without a historical understanding of the development of the concept of the corporation, the best that can be hoped for in curbing corporate harm is piecemeal regulation. However, it would be more effective to redefine the concept, going back to an older view that corporations should be founded for specific purposes which themselves look to the public good.

[35] See French, particularly Part 1, sections 1-2. See also Carl J. Mayer, "Personalizing the Impersonal: Corporations and the Bill of Rights," *Hastings Law Journal*, (March 1990): 577. The current emphasis on regulating corporations has been only partially successful, and, as Grossman argues, understanding the historical development of corporations could suggest more fruitful ways of protecting public health. See Grossman, supra n. 31.

[36] See Russell Mokhiber and Robert Weissman, "Rotten to the Core," *Znet Commentary*, April 8, 2002 for a description of the work of Frank Easterbrook and Daniel Fischel, University of Chicago law professors.

[37] See Richard L. Grossman, "Justice for Sale: Shortchanging the Public Interest for Private Gain," *Defying Corporations, Defining Democracy: A Book of History and Strategy*, ed. Dean Ritz (NY: The Apex Press, 2001). This paper was written in the summer of 2002, as scandals revealing just such attempts continue to make the headlines. It should be noted that perceived interests in the status quo are not always accurate: the record shows both that companies complying with regulation initially thought to be harmful either do not sustain the expected losses, or even benefit from the changes they are forced to make. (See Rampton and Stauber, p. 87.)

[38] See Judis on how the Occupational Safety and Health Agency has been hobbled, and Rampton and Stauber, esp. pp. 26-37.

[39] For an excellent treatment, see Rampton and Stauber, chapter 1.

[40] Steven Vago, *Law & Society*, 6th edition (Upper Saddle River, NJ, 2000), pp. 241-45.

[41] Marshall B. Clinard and Peter C. Yeager, *Corporate Crime* (NY: Free Press, 1980), p. 9; cited in French, p. 341). See also Gilbert Geis, (1978:279; 1994), cited in Vago (p. 241). Not only does corporate crime kill and maim, but it also "Erode the Moral Base of the Law?" (President's Commission on Law Enforcement and Administration of Justice (1967b:104 Bin Vago, p. 241).

[42] Vago, p. 242-3. Historically, fines have been small, prison sentences uncommon (and short or suspended). For a discussion of this problem, see Russell Weissman, "Death on the Job," on the web at http://lists.essential.org/corpfocus/msg00019.html, April 27, 1999; accessed August 15, 2002.

For discussions of attempts to introduce laws against "criminal homicide" that try to change this perception of corporate crime, see *The Safety and Health Practitioner* issues in 1999 and 2000. (Www.safetymags.com). In some cases, white-collar crime is now being taken more seriously. Maximum sentences were lengthened in November, 2001 and some convicted criminals are getting harder punishment.. See Russ Mitchell, "White Collar Criminal? Pack Lightly for Prison," *The New York Times*, Sunday August 11, 2002; from NYTimes.com. Nonetheless, so far from punishing large corporations that violate health and safety laws, the Federal government continues to award them large contracts. See Ken Silverstein, "Unjust Rewards," *Mother Jones*, (May/June 2002): 69-73; 86.

[43] For another interesting example, see Walter Gordon, "Strict Legal Liability, Upper Class Criminality, and the Model Penal Code," *Howard Law Journal*, 26 (1983):781.

[44] For a good introduction, see Rampton and Stauber, chapter nine.

[45] For an interesting introduction to these issues, see Peter Montague, "The Waning Days of Risk Assessment," *Rachel's Environment & Health Weekly #652*, May 27, 1999; *http://www.rachel.org/bulletin/index.cfm?St=4;* accessed August 15, 2002.

[46] Bioethics, as a field, like some of its feeder disciplines, has been slow to accept work that is seriously

critical of the paradigms upon which mainstream work is predicated. For example, although it is now "politically correct" to show awareness of feminist work, its perspectives are rarely incorporated in mainstream work.

[47] To the extent that bioethics is based in philosophy, the political roots of contemporary philosophy are also worthy of inquiry. See John McCumber's *Time in the Ditch: American Philosophy and the McCarthy Era* (Evanston, IL: Northwestern University Press, 2001).

[48] For a more detailed discussion of this issue, see my "Introduction," *Reproducing Persons* (Ithaca, NY: Cornell University, 1996), pp. 9-18, and "Politics and the Curriculum," Robert L. Simon Ed., *Neutrality and the Academic Ethic* (Lanham, Maryland: Rowman & Littlefield, 1994).

[49] For suggestions about how explicitly to connect underlying worldviews and moral discussion of specific issues, see Michael Boylan, *Basic Ethics* (Upper Saddle River, NJ:Prentice Hall, 2000), Introduction and chapter eight.

[50] See, for instance, Ruth Macklin, "Bioethics and Public Policy in the Next Millenium: Presidential Address," *Bioethics* 15.5 & 6(October 2001): 373-81; Steven Miles, "Does American Bioethics Have a Soul?" *Bioethics Examiner* 6.2 (Summer 2002):1,4.

[51] Helen B. Holmes, "Closing the Gaps: An Imperative for Feminist Ethics," in Anne Donchin and Laura M. Purdy, eds. *Embodying Bioethics: Recent Feminist Advances* (Lanham, Maryland: Rowman and Littlefield, 1999).

[52] David Shenk, "Science=Ethics: Problems on Campus," *The Nation*, March 22, 1999; *http://past.thenation.com/cgibin/framizer.cgi?url http://past.thenation.com/issue/990322/0322shenk.shtml*; accessed August 15, 2002.

[53] See Eyel Press and Jennifer Washburn, "The Kept University," *Atlantic Monthly*, March 2000 for an account of recent events at the University of California at Berkeley. See also Jerome P. Kassirer, "Financial Conflict of Interest: An Unresolved Ethical Frontier," *American Journal of Law and Medicine*, 27.2 & 3 (Summer-Fall 2001): 149. He worries that "financial incentives can influence the very questions raised for study, the design of experiments, the interpretation of the findings, and the enrollment of patients," providing compelling examples of each kind of problem.

[54] See the Ottawa Charter on Health Promotion, Health Canada, *http://www.hc-sc.gc.ca/hppb/phdd/docs/charter/*.

[55] For a helpful analysis of the concept of social construction, see Ian Hacking, *The Social Construction of What?* (Cambridge, Mass: Harvard University Press, 1999). There are now significant critical literatures in many fields that consider how assumptions are constructed. See, for example, how feminist literature argues that the position of women, far from being "natural" is constructed. (A nice summary is provided by Nancy Tuana, *The Less Noble Sex* (Bloomington, IN: Indiana University Press, 1993.) For a series of articles that, among other things, uncovers such construction in the realm of law, see David Kairys, *The Politics of Law: A Progressive Critique*, 3rd edition (NY: Basic Books, 1998). These insights are particularly important where legal concepts (such as "corporation") are at issue.

[56] And we have been considering only the more obvious potentially harmful decisions. There is also a need to consider more subtle ones, such as knowingly marketing junk food in large quantities, despite the health consequences. See Mary Duenwald, "An 'Eat More' Message For a Fattened America," *The New York Times*, February 19, 2002 (internet); and Eric Schlosser, *Fast Food Nation: The Dark Side of the All-American Meal* (NY: Perennial, 2002), esp. 239-43.

[57] See Duenwald, Schlosser.

[58] Intriguing studies now strongly suggest that poverty and inequality are central determinants of health and longevity. See R.G. Evans, M. L. Barer, and T. L. Marmor, Eds., *Why Are Some People Healthy and Others Not? The Determinants of Health of Populations* (NY: Aldine de Gruyter, 1994). For a brief discussion of its implications, see Ken Judge, "Beyond Health Care: Attention Should be Directed at the Social Determinants of Ill Health," (editorial) *British Medical Journal*, 309.6967 (Dec. 3, 1999): 1454-55. For useful recent summaries, see Peter Montague's "Major Causes of Ill Health," *Rachel's Environment & Health Weekly #584,* February 5, 1998, and his "Wealth and Health," *Rachel's #654,* June 10, 1999.

[59] Examining this question also requires us to move beyond national boundaries to look at the kinds of international agreements that might be undermining democratic attempts to limit harm imposed by corporations, such as chapter 11 of NAFTA.

DAVID CUMMISKEY

THE RIGHT TO DIE AND
THE RIGHT TO HEALTH CARE

ABSTRACT: Two central issues that shape contemporary medical ethics are, on the one hand, the nature and basis of rights to health care, and, on the other hand, the rights of the dying and the role of the physician when death is imminent. The right to health care and the right to die are typically treated as two distinct sets of issues. These issues, however, are not distinct. As we shall see, if there is a right to die, which I argue there is, then there is also a right to health care. In particular, I argue that respect for the autonomy of persons requires that we respect a person's informed and considered request to hasten death but that concern for persons also supports strong procedural safeguards governing assisted dying. In addition to the normal conditions limiting physician assisted suicide and active euthanasia, I also argue that the right to die provides a justification for a right to health care services.

KEYWORDS: Right to die, Right to health care, Physician assisted suicide, Euthanasia, Double effect, Right to hasten death

Two central issues that shape contemporary medical ethics are, on the one hand, the nature and basis of rights to health care, and, on the other hand, the rights of the dying and the role of the physician when death is imminent. The right to health care and the right to die are typically treated as two distinct sets of issues. These issues, however, are not distinct. As we shall see, if there is a right to die, which I argue there is, then there is also a right to health care.

End of life issues in medical ethics usually focus on how to respect patient autonomy and yet still act with concern for the best interest of the patient; or at least, to act so as not to directly harm the patient. As the ancient oath famously says, above all first do no harm. These are the three core principles of clinical medical ethics: beneficence, non-malificence, and respect for patient autonomy. The ideal patient-physician relationship must honor and sometimes balance these core principles. More specifically, the nature of the right to die and the correlative responsibilities of the physician involve day to day issues that emerge between a particular patient and physician in a clinical setting. Health care justice, by contrast, focuses on the health care delivery system rather than the clinical setting of the patient-physician relationship. The principles in question here are matters of distributive justice, efficiency,

187

M. Boylan (ed.), Public Health Policy and Ethics, 187-202.
© *2004 Kluwer Academic Publishers. Printed in the Netherlands.*

and fairness. This division is, of course, overly simplistic. The health care system is the context in which clinical questions occur, and the system is often a constitutive part of a moral problem or dilemma. Indeed, health care systems may be more or less justified, in part, because of the problems they resolve or cause. In general, if a basic part of the social structure of a society contributes to the violation of fundamental rights, that is a good (prima facie) reason to modify the structure in question.

At present in the United States, the main obstacle to the acceptance of a broader right to control the manner and time of one's death is the combination of unequal access to basic heath care services and the inadequacy of the care of the dying in this country. In the context of the United States health care delivery system, decisions about death may reflect the economic costs of continued care or the inadequacy of palliative and comfort care provided, rather than the deeper values of the patient. Remarkably, rather than recognizing this as a compelling reason to provide both better end of life care and universal access to this care, irrespective of financial considerations, the inadequacy of medical care has been used as a basis to inflict further indignities on dying patients by restricting their right to die as they see fit.

In addition, arguments which focus on inadequate end of life care provide a similar rationale for limiting or prohibiting the now routine decisions to let patients die. The decision to withdraw (often very expensive) life-prolonging care is surely as subject to subtle or explicit coercive pressures as the decision to end one's life when one is not dependent on life support and so must do so by more active means. Yet the opponents of a more expansive right to die (including the AMA) seem to ignore the apparent inconsistency in their position. If lack of access to high quality end of life care is grounds for doubting the autonomy of decisions to actively end one's life, it is also equally an argument against the autonomy of decisions to passively end one's life. The inconsistency here reveals, I believe, an important problem with the current state of thinking about end of life decision making.

In what follows, we first look at the basic rationale for a right to die. More specifically, the right in question is a right to hasten one's death in such a manner that one dies sooner in order to die better. The qualities that make a death a good death or at least a better death are determined by the person's individual conception of the good and also broader religious or spiritual imperatives. The focus is on the character of the death, however, and not simply the decision whether to live or die when the death of person is not already imminent. A right to die, as I conceive it, in no way diminishes the value of each person's life. After looking at the basic rationale for a right to die we turn to familiar objections to taking more active means to hasten death, including physician assisted suicide (PAS) and voluntary active euthanasia (VAE).

Throughout, I argue that pragmatic considerations, and not basic principles, are the primary remaining objections to actively hastening one's death; that these practical objections apply equally to passively hastening death by letting underlying diseases kill people; and I conclude that the appropriate moral and practical response to the practical objections clearly is not prohibition of physician assisted suicide (PAS) and voluntary active euthanasia (VAE), but instead an expanded right to high quality end of life health care services. As a corollary, it also follows that high quality end of life care should be an essential part of basic health care services.

AUTONOMY, BATTERY, AND PUBLIC POLICY

Although reasonable people may disagree about the nature and extent of a right to die, as a matter of public policy in a secular democracy, there is an overwhelming case for a broad and expansive right to die. Clearly, one of the most powerful sources of opposition to a right to die is based in particular religious beliefs and a more general belief that life and death questions are a matter for the divine will and not human choice. Religious doctrines and the judgment of religious authorities are powerful and important determinants of individual moral belief. In a society committed to religious freedom and a doctrine of free faith, however, religious authority does not provide a reasonable basis for public policy or state coercion. Choice about how to die is analogous to the pro-choice position over the decision whether or not to carry a pregnancy to term. The fundamental liberty principle set out by the United States Supreme Court in the *Casey* abortion decision clearly applies equally to end of life decisions:

> It is a promise of the Constitution that there is a realm of personal liberty which the government may not enter ... Men and women of good conscience can disagree about the profound moral and spiritual implications of *actively hastening a death or letting a person die, even by withholding treatment.* Some of us find *assisted suicide and euthanasia* offensive to our most basic principles of morality, but that cannot control our decision. Our obligation is to define the liberty of all, not to mandate our own moral code. The *patient facing death* is subject to anxieties, to physical constraints, to pain that only *he or* she must bear. The suffering of *patients* is too intimate and personal for the state to insist upon its own vision of the *end* of life, however dominant that vision has been in the course of our history and our culture. The destiny *of terminal patients* must be shaped by *their* own conception of *their* spiritual imperatives. (from *Casey* US SC 1992: italicized changes are alteration from the original to apply to end of life decisions. For original

unaltered text see footnote #1)[1]

Although many constitutional principles are controversial, there is indeed a wide-spread overlapping consensus embracing the principle of freedom of conscience, especially on fundamentally personal, spiritual, and religious matters. There is thus a strong presumption in favor of a fundamental right of self-determination at the end of one's life. Furthermore, since, unlike the abortion controversy, the right to die does not involve any issue of the potential life of another, the right of self-determination should be even clearer in this case.

It should be clear to all that respect for autonomy, that is, individual self-deter-mination, provides a clear basis for the individual's right to refuse life prolonging medical care. Some, however, have argued that the right to refuse care is based instead on right to be free from battery and other assaults on one's bodily integrity. This more limited right, it is argued, does not include a right to control the manner and nature of one's death. Persons have a general right to be "let alone" unless they violate the rights of others, and this negative right is the basis for the right to refuse treatment but it is not a sufficient basis, it is claimed, for a positive right to hasten one's death.[2] This argument is puzzling. Granted that battery is wrong, but so too is restricting liberty. Persons have a fundamental interest in determining how they die, and this provides a clear basis for a basic right to hasten death.[3] We need some com-peting essential government purpose for any state actions that aim to restrict this lib-erty, and this includes attempts to restrict the liberty of others who are willing to help hasten my death. Both sides of this debate recognize the importance and centrality to a person of the manner of their own death. Indeed the debate is over how one shows due respect and compassion for the dying person. So, as an issue of public policy, limits on the right to die need to focus on legitimate protections of the dying person's own interest, or alternatively on distinct and assignable harm to others in society. The rest of our discussion will focus on these types of concerns.

THE PRIORITY OF PERSONS OVER MERE PREFERENCES

First, we need to get clearer about the nature of respect for a dying person. It is argued by some that the intentional destruction of an innocent person is always incompatible with respect for the dignity of that person. Indeed, the argument goes, it is always wrong to kill oneself merely to avoid suffering because one's value and dignity as a person transcends the mere value of pleasure and pain. It is thus never permissible to kill oneself because one's future promises more misery than benefit because doing so violates the fundamental moral requirement to treat oneself as an end and not a means only. Each person has a special value, a dignity that cannot be exchanged for mere benefit or to avoid some harm. Just as it is wrong to sacrifice someone in order to promote the pleasure of others, so too it is wrong to sacrifice oneself merely to avoid pain. The self-destructive act of suicide treats oneself as a thing and thus fails to recognize the inner value and dignity of oneself as a person.[4]

The idea behind this principle of prohibition of all suicides does have some appeal. Consider a case where a person whom one cares about is acting on self-destructive preferences: Normally, if we care about people, then we also care about their happiness and we thus also want them to realize their projects and to live up to their ideals. Our concern for the ends of others, however, is based on our concern for them. When we act in ways that promote the ends of another, it is because we think that that person matters. If not, then their ends would not matter. So there is something incoherent about caring about the subjective ends of another and not caring about that person. In general, the ends of another person matter because the person matters. The value of a person is prior to the value of the person's preferences — so far so good.

The argument for the prohibition on suicide next assumes that one's continued existence as a person always takes priority over one's conception of one's own good. Here is the idea: My conception of the good matters only if I matter, so my conception of the good cannot provide a justifying basis for destroying myself. Suicide for the sake of benefit is thus never justified. Despite its apparent logic, this just does not seem right. It seems clear that death can be a release and a benefit for a dying person. So what has gone wrong?

We need to distinguish two senses in which we might be concerned for a person, and not simply the preferences of the person. The argument for prohibition assumes that concern for a person necessarily involves concern for the continued existence of a person. Concern for a person, however, involves concern for the person's integrity and character, not the person's mere existence. The object of respect and the basis of human dignity is autonomy: the capacity to set oneself ends and pursue a conception of the good that gives one's life meaning and purpose. We thus respect human dignity by endorsing and following procedures that allow each person to reflectively endorse and to pursue a substantive conception of the good.[5] We respect the dignity of ourselves by living *and dying* in accordance with the values and principles that we reflectively endorse. Respect for others also involves a similar respect for their values and principles. And, of course, respect for dignity, in this sense, may include assisting them in dying, in a manner that reflects the values and principles that constitute their conception of the good. Helping someone die may be fully consistent with respecting the dignity of that person.[6]

The choice here can be limited to how to die when death (or loss of personhood) is imminent. Controlling the timing and method of one's death in no way compromises the value of life itself. One can continue to view life as a priceless gift and to honor the dignity of all persons, including one's self, and yet also realize that one's life as a person is now done. The acceptance of death is compatible with the valuing of life. After all, it is not as if how long one lives determines the intrinsic value of one's life. Furthermore, I believe that clinging to every minute of life, simply because it is more life, fundamentally misconceives the nature of the intrinsic value of a human life.

Let us pause for a moment and notice, however, that the argument so far does not imply that one should honor all requests to die. The decision to die must be informed and considered and it must actually reflect the values and principles of the person in question. Respect for a person does not require respect for uninformed or reckless or irrational decision. I have argued that the primary object of respect is the person and not simply the person's preferences. We honor preferences because they reflect the character and values of the person. Especially in cases where the consequences are serious and irreversible, it is reasonable to have procedures and policies that help demonstrate that the preference in question is indeed informed and that it reflects the values and principles of the person. The decision to die clearly should be based on a stable and enduring principled preference. It follows that safeguards that try to insure that the decision to die does indeed reflect the person's values and principles are called for. We want to make sure that a request to die sooner is not simply the result of depression, panic, or coercive social pressure. Procedural safeguards reflect a commitment to respect the person, and not just passing or distorted preferences, and thus these types of safeguards are not unduly paternalistic and disrespectful of the person. The exact nature of these safeguards will be discussed more explicitly below.

THE ROLE OF THE PHYSICIAN

Assisting someone to die strikes many physicians as contrary to their fundamental role as healers. The role of the physician is to save lives, not end them, it is argued.[7] Some physicians (still) think of death as an enemy, an evil foe to be conquered and defeated and, since this is in the end impossible, to be fought by every means necessary to that bitter end. Of course, few physicians really hold this extreme view (although the above is almost a direct quote from a physician on a medical ethics committee). Patients clearly cannot always be cured and the physician's role includes caring for patients who cannot be cured. Indeed, before the great successes of modern medicine, easing the suffering of illness and dying was a primary focus of much medical care.

Yet once we acknowledge, as surely one must, that caring for the dying is a constitutive part of the role of a physician, it is unclear why helping someone die cannot be an act of care that is done for the sake of the patient in question. Since the fundamental values and principles of a particular patient determine what is a good death for that patient, the patient's values, not the physician's, should be sovereign. Since caring for a person includes respecting their values and perspective, helping a patient to die sooner so that they die better can be a way of caring and showing compassion for the dying. Compassion towards the dying is indeed a medical imperative, and compassion in dying in can involve hastening the death of the person. When this is so, there is nothing in the role of the physician that is intrinsically inconsistent with helping a person die in a manner that reflects the dying person's values and perspective.

Furthermore, physicians do let patients die, in a compassionate and caring way, all the time. It is now commonplace to withdraw life-sustaining treatments at the request of patients or their surrogates. Clearly, when such decisions are made, the physician role does not end; it is instead transformed. The standard of reasonable care of the dying includes continuing to care for the patient, to respect the wishes and values of the patient, and striving to make their passing as comfortable and painless as is possible. Why then does this role not also include respecting a terminally ill patient's considered and informed request to hasten death? In short, hastening a patient death is inconsistent with the role of the physician only if it is wrong for some other reason. It is a dodge, and ultimately also disingenuous, to appeal to the "essential healing" role of the physician. Caring for those who cannot be cured is and ought to be part of the mission of medicine.

The substantive objection behind this misleading appeal to the ends of medicine, I suspect, is the concern that it is always wrong to intentionally end life. What is truly difficult in these controversial cases is quite simply the individual's recognition of the weighty responsibility for helping to actively end a life. This is not a decision or action to be taken lightly. In the Netherlands, whenever possible, physicians always try to ease a patient's death by euthanasia at the end of the day and on a Friday.[8] It is too hard to go on as usual afterwards. Even with the conviction that the action is compassionate and justified, hastening a death is still a trying and difficult thing to do. It is no wonder that physicians do not seek this weighty responsibility.[9]

INTENDING AND FORESEEING DEATH

At the heart of much opposition to actively hastening death is the assumption that the intentional taking of human life always violates the sanctity of life. This type of concern is not new. Over the past 40 years, the progress of medical science and technology has forced us to continually rethink and consider the meaning of the prohibition on taking human life. First, the ability to keep the body functioning when the brain is dead changed the focus from cardio-respiratory function to brain function in the determination of the death of the person. Next, the New Jersey Supreme Court case that involved Karen Anne Quinlan (in 1976) focused attention on the permissibility of withdrawing life support for persons in a persistent coma with irreversible loss of consciousness. Although at the time the hospital and doctors argued that the act of withdrawing life-sustaining care is an intentional action that causes death and so is an act of wrongful killing, it is now almost universally agreed that such actions are not wrong. It is also argued that, in withdrawing life-preserving treatment, the physician is not killing the patient but is simply letting the patient die. It is the underlying disease, and not the physician, they say, that causes the patient's death. Although this claim is a commonplace, it is curious. As we have just discussed, the opponents of actively hastening death have also argued that curing disease and preserving life is the primary role of the physician. We are thus owed some explanation for why it is ok to just let patients die when life-sustaining care is available. Of course, if the patient is not facing a terminal condition and has not consented, it would be truly surprising if this omission were consistent with the ends of medicine and the role of the physician! It seems clear that what distinguishes permissible and impermissible cases of letting someone die is in part the consent of the patient or a surrogate, or, when the patient's wishes are not known, a judgment that the continued life prolonging treatment is not in the patient's best interest. The patient's consent, fundamental values, and best interest play a crucial role in deciding whether or not continued life-prolonging care is called for.

So, what importance, if any, is there to the fact that the disease is necessary in these cases as the underlying cause of death? The opponents of more active measures to hasten death, when the death is equally consistent with the patient's wishes, fundamental values, and best interest, place tremendous importance on whether the underlying disease kills the patient or whether instead an additional cause is introduced to intentionally cause the death. Clearly many people *feel* that there is a personal responsibility for the death if more active measures are taken, even when the death is just as certain a result in both cases. But what is the morally relevant difference between intentionally withholding or withdrawing care that will result in the death of the patient and intentionally introducing a cause that will result in the patient's death? Why is the former a case of permissibly letting a patient die and the later a case of (supposedly) impermissibly killing the patient?

The typical, but mistaken, answer to this question appeals to the supposed different intentions in the two types of cases.[10] The moral principle that is supposed to mark the difference here is called the doctrine of double effect. Roughly, this principle distinguishes actions that directly aim at harm from actions that cause, as a foreseeable effect, a similar harm but which do not directly aim to cause harm. Fortunately, we do not need to consider the soundness of this controversial principle, for it simply does not even apply to the types of cases that we are considering.[11]

The doctrine of double effect prohibits intentionally aiming at evil (or harm to the innocent) so that good may be done. The principle is familiar in discussion of just war theory and terrorism. One way of distinguishing acts of terrorism focuses on the wrongness of directly harming or killing the innocent as a means to even an otherwise legitimate goal. On the other hand when in war, for example, one may foresee that innocents will be killed as a result of an otherwise justified bombing of a military target. Foreseeing that innocent will be killed is morally different than targeting the innocent. It makes a difference if one is aiming at the harm to innocents, that is part of the plan one might say, or whether it is an unintended result of what one intends. Although the principle itself is controversial, let's assume that it is sound. The problem is that this principle simply does not apply to the types of cases which we are interested in here. The doctrine of double effect presupposes that the patient is being harmed and seeks a context in which the outcome is not a wrongful harm. Yet the reason why it is permissible to intentionally let a patient die, in some situations, is that we are respecting the patient's considered preference or fundamental values, and thus not letting the patient be harmed (all things considered) at all — or, alternatively, the death in question is judged to be in the patient's best interest and thus simply is not a harm. As these judgments about the permissibility of letting a patient die concede, death is not always an evil or harm to a person. In some cases the body lives on but the person is already lost and thus intentionally acting in ways that are meant to let the body die manifest no intention to allow a harm at all. In other cases, death can be a release from great suffering and thus something the person legitimately deems good. Indeed, if letting a person die involves foreseeing that the patient will be harmed, it is not clear *why* the physician does not do wrong by letting patients die. At the very least, if the death is a harm, the physician should do all they can to discourage patients from intentionally harming themselves in this way. Clearly, it is hard to see how it could ever be permissible for a proxy to harm a patient in this way. And of course, the best interest standard of proxy decision-making would simply never apply. The point here, however, is that we rightly judge that death is not always a harm. But if the patient is not really harmed, then so too actively hastening death also does not necessarily intentionally harm the patient. In point of fact, the doctrine of double effect, as used in these types of cases, illegitimately assumes what is in question — namely, that the act of ending a life in accordance with the considered wishes of the patient is always wrong and thus evil. It assumes what is supposed to be shown and thus does no independent work in distinguishing actively hastening death from letting a person die.

As a last point, we need to briefly consider the case of "terminal sedation." As a result of Cruzan (1990), it is permissible to withdraw nutrition and hydration and sedate a patient and let them die. This practice of "terminal sedation" is now also common in the United Kingdom and many other countries. How the doctrine of double effect here applies is especially curious. Although the sedation does indeed often treat pain and suffering, the lack of nutrition and hydration serves no palliative function. The reason for the withdrawal of nutrition and hydration is to let the patient die. This is the end and goal just as clearly as in the case of a lethal injection. No additional causal agent is introduced and thus the process is slow rather than swift, but the end is just as certain. One hopes that this type of decision is not treated lightly; that safeguards would be in place to prevent abuse; and that all other options are first explored. Yet I fear that a clear negative side effect of the comforting use of the idea that it is an underlying disease, and not the physician, causing the death is that decisions to end a patient's life by withdrawal of life prolonging care are not adequately scrutinized and given the attention that they deserve.

LIMITS AND SAFEGUARDS

The principles of respect and concern that I have been defending are focused on a person's *informed* preferences that *reflect the person's basic values*. Decisions at the end of life, however, are difficult decisions influenced by fear of suffering, fear of death, and perhaps even clinical depression. Patients also often have unrealistic fears and concerns about loss of independence. It is thus necessary to be cautious in responding to a person's desire to die. We need to know if the expressed preference to die is a considered, informed preference that reflects the person's values or whether it is a suicidal impulse, which simply reflects fear and depression. To overcome these difficulties, yet still respect the right to choose death, assisted dying statutes in Oregon (and proposed statutes in other states), include mandatory waiting periods, second opinions, counseling about palliative care and hospice care options, possible psychological evaluation for clinical depression, and a provision to try to include family members in the discussion and thereby get a fuller sense of perspective on the patient's preferences.   If one includes appropriate *safeguards*, then I believe that the benefits of honoring basic rights overrides these otherwise legitimate concerns about hastening death.

The standard conditions and safeguards that are now incorporated into statutes permitting actively assisted dying, and which I would support, typically include:

(1) A voluntary request initiated by a competent individual
(2) A mutual, informed decision-making process
     — including an understanding of the reason for the request
(3) A critical and probing consideration of alternatives
     — including curative, comfort, hospice, and palliative care options
(4) Consultation with others
     — including an independent physician, perhaps a psychological or psychiatric consult, and perhaps family consultation.
(5) A continued, expressed, preference for death
     — including an explicit written request and a mandatory waiting period
(6) An irreversible condition causing the permanent loss of self
     — terminal illness
     — perhaps, and more controversially, significant dementia

These procedural safeguards clearly can have an impact. In the Netherlands, although about 25,000 patients per year seek assurances of the option of an assisted death and 9,000 patients make an explicit request, only 2,320 result in active euthanasia and 400 choose assisted suicide. In the state of Oregon, palliative care and hospice use has improved significantly since the legalization of assisted suicide. It is likely that this is the result of a more open system that allows patients to speak more freely with their physicians and thus become better informed about their options. Nonetheless, even with excellent palliative and hospice care, some patients still want the options to hasten their death, if their condition becomes intolerable.

It is important to realize that physical pain alone is rarely given as the only reason for wanting the option to hasten one's death. Even with high quality palliative care, many patients still want to control the manner and time of death. Therefore, it is a consequence of a prohibition on assisted dying that we use the force of law in an attempt to compel fully competent informed adults to die in a manner that is contrary to their values and principles. This is not something to be done lightly. Imposing, in a coercive and paternalistic fashion, one's own conception of how one should die on another person is fundamentally disrespectful and a basic affront to the dignity of that person. The opponent of assisted dying chooses to force people to die in a manner that offends the person's preferences and principles.

We can, of course, foresee that even with safeguards some people will choose to die for irrational reasons. We must thus decide how we should balance rights of self-determination and duties to prevent unintended and indeterminate harm. In answering this question, the doctrine of double effect provides guidance. Other things equal, it is wrong to intentionally infringe on the rights of some because we foresee that other persons may be harmed. The proper response is instead to honor individual rights and to also strive to minimize any foreseeable harm. Indeed, the procedural safeguards outlined above aim to balance respect for individual self-determination with our legitimate concern to protect the vulnerable.[12] There is a big difference, however, between procedures that aim to promote an informed, voluntary choice and outright prohibitions that paternalistically assume that dying patients are uninformed or incompetent.

THE RIGHT TO DIE AS A BASIS FOR HEALTH CARE RIGHTS

Many people still oppose any form of active assistance in dying in the United States simply because of the unequal access to high quality end of life care in the United States. This is a bad argument despite its popularity. First, the conclusion that follows from concerns about inequalities in access to health care is that we should be fighting for universal access to a basic health care package that includes high quality end of life care. One can only wonder at those who have opposed and continue to oppose universal coverage but who then use the lack of universal coverage to deny people basic rights to self-determination at the end of life. Second, if I am faced with a system that has inadequate end of life care, then it may well be the case that I prefer to die sooner in order to die better. That I would not have this preference if there were better end of life care available in no way undermines the soundness of my preference given the real life options that I face. Surely I would also not choose to die if presented with a cure for my disease, but fantasy options do not change my considered preference given my actual situation. Third, why would the fact that I am denied basic rights to health care services provide a reason for also infringing on my basic right to control the manner of my own death? It is strange logic indeed that uses the violation of one right as the basis for the violation of another right.

The proper response to the inadequacy of health care services is clearly to strive to provide better health care services. Furthermore, as a general principle if right B is necessary to secure a fundamental right of type A, that is a clear basis for protecting right B. For example, the right of free association (right B) is a necessary social precondition for the right to assemble and petition one's government for grievances (right A). Even though only the right of assembly is enumerated in the Constitution, the right of association is also a constitutionally guaranteed right simply because it is necessary to protect the enumerated right of assembly. Since we have a fundamental liberty right of self determination in dying (right A), and if a right to health care services (right B) is a social precondition for the safe exercise of this fundamental liberty right (right A), then we have a social responsibility to provide health care services (right B), especially high quality end of life care. The fundamental right to die is thus itself a basis for recognizing a right of universal access to basic health care services.

Finally, opponents of assisted dying need to explain why the allegedly coercive context of US medicine does not provide an equally sound argument for prohibiting physicians from withholding and withdrawing care and "letting people die." The economic and social pressures here are as great, and usually in fact much greater, than in the case of patients that are not life-support dependent. Clearly, there will be countless cases where the withdrawal of care from a patient in an ICU is many times more cost effective than the cost of home care or hospice care for a cancer or AIDS patient. The argument from social pressure and economics is relevant to all end of life medical decisions. We should be more concerned with insuring that the decision to die, whether it involves passively letting someone die or more actively hastening death, is fully informed and truly voluntary. Given the clear inequities in our health care system, for millions of people the decision to die must surely be affected by the economic consequences of the decision. So here again we have an overwhelming reason to provide universal access to high quality health care.

CONCLUSION

The right to die and the right to health care are indeed connected. Contrary to the arguments of many, however, the connection does not provide a basis for limiting the right to die to passive cases of withholding and withdrawing care and thereby letting patients die. First, pragmatic arguments that focus on the coercive force of socio-economic considerations apply equally to all end of life decisions. Second, the proper conclusion is not that there is no right to actively hasten death. Instead, the right to die provides an additional reason (as if we do not already have enough) for universal access to health care services. Finally, as long as millions are denied the right to health care, the less risky practice of physician-assisted suicide increases the self-determination of competent patients facing death without undue risk. If we are to show due respect for patients facing death, we should show them concern and compassion but the final decision about how to die must be theirs.

NOTES

[1] Italics indicate passages altered above so as to apply to the end of life cases: "It is a promise of the Constitution that there is a realm of personal liberty which the government may not enter ... Men and women of good conscience can disagree about the profound moral and spiritual implications of *terminating a pregnancy, even in its earliest stage*. Some of us find *abortion* offensive to our most basic principles of morality, but that cannot control our decision. Our obligation is to define the liberty of all, not to mandate our own moral code ... The *mother who carries a child to full term* is subject to anxieties, to physical constraints, to pain that *only she* must bear. *Her* suffering is too intimate and personal for the state to insist upon its own vision of the *woman's role*, however dominant that vision has been in the course of our history and our culture. The destiny of the *woman* must be shaped ... by *her* own conception of *her* spiritual imperatives."   (from *"Casey v. Planned Parenthood of Pennsylvania"* US SC 1992)

   The constitutional argument for the right to hasten death is developed in "The Philosopher's Brief," by Ronald Dworkin, et al., *New York Review of Books*, 44.5 (March 27, 1997).

[2] This commonplace argument can be found in the influential New York Task Force on Life and Law, "*When Death Is Sought: Assisted Suicide and Euthanasia in the Medical Context*," Supplement to Report, April 1997. For a good response to this argument, see G. Dworkin p.69 *in Euthanasia and Physician Assisted Suicide: For And Against*, Gerald Dworkin, R.G. Frey, and Sissela Bok (Cambridge: Cambridge University Press, 1998).

[3] An fundamental liberties based argument for the right to die is developed in "The Philosopher's Brief," which is an amicus curiae brief of six moral philosophers to the United States Supreme Court pertaining to the cases of the *State of Washington v. Glulcksberg* and *Vacco v. Quill*. For the main idea, see Dworkin's "Assisted Suicide: The Philosopher's Brief, Introduction" in *The New York Review of Books*, 44. 5 (March 27,1997).

[4] Kant's example of a suicide maxim in the *Grounding for the Metaphysics of Morals* is often taken to support this type of position. Kant's discussion of maxims of suicide in the *Metaphysics of Morals*, however, is subtler. Leon Kass asserts this type of position in "Neither for Love or Money: Why Doctors Must Not Kill," *The Public Interest* 94 (Winter 1989). The best fleshing out of this position is found in David Velleman's article, "A Right to Self Termination," *Ethics* 109.3 (April 1999): 606-28.

[5] The understanding of respect for persons and human dignity here is, of course, heavily influenced by John Rawls.

[6] In the quite different context of his defense of the death penalty, Kant recognizes that respect for human dignity does not always require that we preserve life. So in fact he does not unequivocally endorse a substantive interpretation.

[7] This is the position of the American Medical Association and the British Medical Association. It is also defended by Leon Kass in "Neither for Love or Money: Why Doctors Must Not Kill," *The Public*

Interest, .94 (Winter 1989), and in "Is There a Right to Die?" *Hasting Center Report,* (Jan 1993): 34-43. Also see Edmund Pelligrino "Doctors Must Not Kill" in *The Journal of Clinical Ethics,* (Summer 1992): 95-102. Despite the position of the medical associations, this does not seem to be the position of many and perhaps most doctors. Many doctors support physician assisted suicide and many who oppose it do so for the more pragmatic policy reasons discussed below. For a quite different view of the role of a compassionate and responsive physician, see Timothy Quill, "Doctor I want to Die. Will you help me?" *Journal of the American Medical Association,* (August 1993): 870-73, and *Death and Dignity: Making Choices and Taking Charge* (Norton, 1993). Also see Gerald Dworkin and R. G. Frey excellence response to Kass in *Euthanasia and Physician Assisted Suicide: For And Against,* Dworkin, Frey, and Bok (Cambridge, 1998).

[8] Liesbeth (Elisabeth) Kalff, NVVE, The Dutch Voluntary Euthanasia Society, presentation on euthanasia in the Netherlands, Amsterdam, March 2003.

[9] There are also conceptual limits on how we think about hastening death. Margaret Battin has pointed out the interesting difference in this respect between attitudes towards hastening death in Germany and the United States (see her article, "Euthanasia: The Way We Do It, The Way They Do It," Journal of Pain and Symptom Management, vol. 6 no.5, 1991; pages 298-305). She argues that German cultural history and linguistic resources have given root to a somewhat unique position on physician-assisted "rational suicide." First, the German language, and thus also thought, distinguished different kinds of acts we call suicides with different words which carry distinct denotations and connotations. The German *Selbstmord* means "self-murder" which connotes a desperate and wrong action; *Selbsttotung* literally means self-killing and is a neutral scientific term; the Latin construction *Suizid* is used for a suicide which results from psychiatric pathology; and *Freitod* means "free death" and connotes a voluntary individual choice which is an expression of deeply held values or ideals. One commits *Selbstmord* but one chooses *Freitod.* These distinct linguistic categories have made it comparatively easy for Germans to distinguish a free death from unacceptable suicide. The United States on the other hand, has struggled with the confusion generated by using one word for similar actions motivated by such different contexts and reasons. Second, in Germany, a "freely chosen death" or rational suicide — that is, a voluntary, reflective choice to end one's life, which is not rooted in despair, clinical depression, or mental illness — has been decriminalized since 1751. So in addition to a richer linguistic conceptual scheme, the evolution of German practice has been unencumbered by either a legal prohibition or an unclear common law traditions.

As a result, medically related assisted "suicide" is quite common in Germany. The German Society for Humane Dying (which is similar to the Hemlock Society) lends out a booklet under control condition on "A Dignified and Responsible Death" which details how to acquire and take drugs so as to produce a painless and nonviolent death. The society reports between 2,000 and 3,000 suicides a year among its more than 50,000 members (as of 1991). The problem with the German approach is that with no physician involvement or statutory procedural safeguards one is likely to have a significant number of cases where the patient in question is not adequately informed about other medical options. In addition, since initial requests to die are often a result of unwarranted fear or treatable clinical depression, physicians could play an important role in ensuring that patients requesting death are making an informed, reflective, and uncoerced decision.

[10] See, for example, Edmund D. Pellegrino, "The Place of Intention in the Moral assessment of Assisted Suicide and Active Euthanasia," in *Intending Death: The Ethics of Assisted Suicide and Euthanasia,*

edited by Tom L. Beauchamp (Englewood Cliffs, NJ: Prentice Hall, 1996).

[11] For the best discussion of the doctrine of double effect and its problems, see Shelly Kagan, *The Limits of Morality* (London: Oxford, 1991). Also see Timothy E. Quill, M.D., "The Ambiguity of Clinical Intentions," in *The New England Journal of Medicine*, 329.14, (September 1993): 1039-40.

[12] Can a policy of assisted dying provide adequate safeguards against involuntary euthanasia? The Remmelink study has suggested to some that an unacceptable level of misuse is unavoidable. What the Dutch classify as involuntary euthanasia occurred in 0.8% of all deaths. Somewhat triumphantly, many have announced that this is powerful evidence that significant abuse is impossible to avoid once a policy of allowing euthanasia is set in place. This conclusion, however, is overly hasty.

First, voluntary euthanasia in the Netherlands is taken to require an explicit, informed, and persistent request to die. All of the cases without an explicit request involved mentally incompetent patients. On average, the specialist involved in these cases knew their patients for 2.4 years and the general practitioners for 7.2 years. In 60% of the "involuntary" cases the physician had clear evidence of the patient's preferences for death, either from earlier discussions or from family members. In 83% of these controversial cases, the physician had the consent of a relative. In the remaining cases, the patients were incompetent and suffering from uncontrollable convulsions and significant pain and life was probably shortened by a few hours or days at most.

The physicians in these cases were attempting to act in the best interest of their patients and agonized over their decisions. This is a far cry from the reports of physicians killing patients against their will that have been reported by overzealous critics. In the United States all such surrogate decision-making is considered a natural extension of autonomy and it is now routine in cases of passive euthanasia. Once we distinguish Voluntary, Non-Voluntary (no explicit request), and Involuntary (contrary to known preferences of the patient) cases, all of the problematic cases are properly classified as non-voluntary – there is no evidence in any of these cases that the action was contrary to the preferences or principles of the patient. By US standards, these are all cases of legitimate proxy judgment by a surrogate acting on a prior directive (even if not an explicit specific request), or trying to determine the preferences and best interest of the patient. (see L. Pijnenborg, van der Mass, et al., "Life Termination Acts Without Explicit Request of Patient," *The Lancet* 341 (1993): 1196-99.)

The Netherlands seem to demonstrate that it is indeed possible to practice institutionalized voluntary euthanasia without sliding down a slippery slope to either euthanasia on demand or the involuntary killing of the sick and the elderly. We also learn that the healing ethic of the medical profession is alive and well in the Netherlands.

[13] I would like to thank Sarah Conly, Tom Tracy, Michael Boylan, and the students of my Colby, Bates, and Bowdoin comparative medical ethics and justice program for helpful comments and discussions.

MICHAEL J. GREEN

GLOBAL JUSTICE AND HEALTH:
IS HEALTH CARE A BASIC RIGHT?

ABSTRACT: This paper concerns whether health care is a human right. Specifically, it discusses Henry Shue's claim that it is what he calls a basic right. Shue argues that security rights, such as rights against physical assault, and subsistence rights, such as rights to health and food, are basic rights insofar as they are necessary for the enjoyment of other rights. This apparent necessary connection provides an argument for basic rights that, it seems, even those with minimal commitments to rights must accept. The paper discusses this argument for basic rights and its connection with Shue's claim that all basic rights have three kinds of corresponding duty. It ends by considering whether justice and human rights might come apart, specifically, in cases where the protection of human rights seems to involve unfair demands on others.

KEY WORDS: Global Healthcare Justice, Henry Shue, Basic Rights, Correlative duties, International ethical duties.

Are international inequalities in health care unjust? It is well known that the gap between those who receive the best health care in the world and those who receive the worst is staggering and that which side of the gap one is on makes a significant difference in the length and quality of one's life. These inequalities certainly seem unfair in the sense that they are important and cannot plausibly be said to be deserved or chosen by those who lack decent health care. But are they unjust? That is, are those individuals or societies that could rectify these inequalities unjust because they fail to do so? Is the world's economic and social system unjust because it contains them?

There are two hurdles to declaring that these inequalities are unjust. First, they concern health care which is something that some people must take affirmative steps to provide to those who need it. It is more controversial to say that inequalities that are allowed to persist are unjust than it is to say that inequalities due to avoidable behavior are. Second, the inequalities in question are international. The claim that an unequal distribution of resources across different societies is unjust is more controversial than the claim that inequality within a particular society is unjust, even when what is unequally distributed is a vital resource such as health care.

Perhaps there is a way of cutting through both problems at the same time. If health care is a human right, then, it seems, its unequal provision would be a matter of jus-

M. Boylan (ed.), Public Health Policy and Ethics, 203-221.

tice. Of course, most societies treat health care as an important goal of public policy. But showing that health care is a right would show that it has special significance. Rights, it is believed, take priority over other goals and cannot simply be balanced against other ways of promoting the general welfare; they can be sacrificed only when the cost of respecting them would be extremely great. Furthermore, if health care is a right, then we will have to employ a special standard for assessing public health policy. It will not be enough to improve aggregate or average health in a society but rather the health needs of each individual will have to be addressed since, by hypothesis, each individual has a right to health care. In assessing a health policy, in other words, we cannot look just at overall improvements or declines but will have to attend to the distribution of health care across the population that is covered by the policy as well. Finally, showing that health care is a right strongly suggests an answer to my question about justice insofar as the systematically unequal protection of human rights is a good candidate for injustice.[1] Showing that health care is a human right, moreover, would show that it is a matter of international concern making it very likely that we will be forced to conclude that the world is seriously unjust.

In this paper, I will take up one attempt to show that health care is a human right: Henry Shue's argument that a minimal level of health care is what he calls a basic right, that is, one that is entailed by all other rights. After presenting his conception of rights, I will examine his two main theses about basic rights: that they are entailed by all other rights and that they have at least three kinds of corresponding duties. In the last section, I will return to the connection between justice and rights. Of course, there are many conceptions of rights and, by limiting my attention to this one, I will not definitively settle the relationship between justice, rights, and health care. However, I believe that Shue's discussion of basic rights is both deeply interesting on its own and that it can advance our understanding of the broader issue of the connection between justice and rights.

RIGHTS

A moral right, according to Shue, "provides (1) the rational basis for a justified demand (2) that the actual enjoyment of a substance be (3) socially guaranteed against standard threats."[2] The *substance* of a right is what the right is a right to.[3] For example, I enjoy the substance of a right to enough food to live if and only if I have enough food to live. Those who hold rights can make demands on others but one cannot straightforwardly demand that others provide or protect the substance of one's rights. What one can demand is that others provide social guarantees against standard threats to the enjoyment of the substance of one's rights where standard threats are common, prevent or severely hinder people from enjoying the substance of their rights, and can be remedied or eradicated by social arrangements.[4] To say the same thing in other words, the duties correlative to rights are to provide social guarantees against the standard threats to the enjoyment of rights.[5]

Shue claims that some rights are *basic rights*, meaning that their enjoyment is "essential to the enjoyment of all other rights."[6] He maintains that at least three broad classes of rights fit this definition: rights to physical security, such as rights against murder, torture, or assault; subsistence rights, such as rights to food, clothing, shelter, and health care; and rights to liberty. I will consider two of the theses that Shue advances about basic rights. The first thesis is that the existence of basic rights is entailed by the existence of any other right. Because the enjoyment of basic rights is necessary for the enjoyment of all other rights, it appears that anyone who believes that some rights exist is committed to accepting the existence of basic rights as well. Thus Shue proposes the following argument for basic rights.

1. Everyone has a right to something.
2. Some other things are necessary for enjoying the first thing as a right, whatever the first thing is.
3. Therefore, everyone also has rights to the other things that are necessary for enjoying the first as a right.[7]

The first premise of this argument is assumed. The argument is not meant to refute doubts about rights in general but is addressed to those who believe there are some rights and have doubts about basic rights. Specifically, Shue is concerned to rebut those who accept civil and political rights but doubt either that social and economic rights exist at all or that, if they do exist, they have the same priority as civil and political rights.[8] While there is no canonical list of civil and political rights, they are generally thought to include rights to personal liberty, security, privacy, thought, expression, religion, and assembly and rights against torture, slavery, and inhumane punishment.[9] The success of the argument can be measured by its ability to show that someone who accepts a reasonable part of this list is thereby committed to accepting basic rights.

The second thesis is that there are at least three kinds of duties corresponding to all basic rights: *deprivation duties,* duties not to deprive others of the substance of their basic rights; *protection duties,* duties to protect others against deprivation; and *aid duties,* duties to aid those who have been deprived.[10] The tripartite analysis of duties is an expository convenience and does not commit Shue to holding that there are only three kinds of correlative duties and no more. On the contrary, he endorses

Jeremy Waldron's more open-ended formulation that rights give rise to an indeterminate number of "successive waves of duties."[11] The reason for thinking that basic rights have many corresponding duties goes roughly like this. Moral rights protect individual interests by imposing moral duties on others. But there need not be only one duty for every basic right. Since the point of basic rights is to protect each person's access to substances, a basic right may have as many correlative duties as are necessary to guarantee access. Since there are typically many different threats to access, there are many different duties to protect access in turn. In particular, it makes sense to have duties to make up for failures. Since it is overwhelmingly likely that people will violate one another's rights, the protection of interests has to include protection against others.

Shue's strategy for combating the priority of political and civil rights over social and economic ones involves three points. First, he argues that a subset of political and civil rights are basic rights. This subset includes rights against murder, torture, mayhem, rape, or assault; Shue refers to these rights as *security rights*.[12] Second, he argues that what he calls *subsistence rights* are also basic rights, where subsistence rights are rights to whatever is needed for "a decent chance at a reasonably healthy and active life of more or less normal length" including "unpolluted air, unpolluted water, adequate food, adequate clothing, adequate shelter, and minimal preventive public health care."[13] The symmetry of the arguments for the relatively uncontroversial security rights and for the more controversial subsistence rights undermines the claim that only the former are genuine rights.[14] Finally, Shue argues that both security and subsistence rights have negative and positive corresponding duties; this undermines the objection that political and civil rights take priority over social and economic ones on the grounds that the former are negative rights while the latter are positive rights.

The strategy is indirect. It does not rest the case for subsistence rights on the importance of the interests that they protect. Rather, the argument for subsistence rights is based on the claim that they are entailed by other rights. As such, the argument seems to draw very strong conclusions from quite modest premises: anyone who believes some rights exist is committed to accepting subsistence rights as well. Nor does the argument rely on any particular account of the basis of rights. For example, it does not depend on showing that the values underlying civil and political rights also provide a rationale for subsistence rights. Rather, the argument appears to rely only on what is involved in thinking that there are any rights at all. Needless to say, it would be a significant and surprising achievement to derive this conclusion from such weak premises. The possibility of doing so, in any event, will be my main concern. I will raise some questions about the indirect argument for subsistence rights. In doing so,

I do not mean to deny that these rights exist or that a more direct strategy of arguing for them would succeed. In fact, I agree with a theme that Shue forcefully and persuasively emphasizes: since the interests protected by subsistence rights are at least as important as those protected by security rights, it is hard to believe that there is a rationale for a system of rights that primarily or exclusively protects the latter. Nonetheless, I will focus on the indirect argument outlined above because it seems to demonstrate that there is a surprisingly wide and rarely appreciated commitment to subsistence rights and I am curious about how far this specific argument will take us.

BASIC RIGHTS

The argument for basic rights is that they are necessary for the enjoyment of all other rights. If so, it seems that those who believe that some rights exist are committed to acknowledging that basic rights exist as well. This section concerns how basic rights are necessary for other rights. Shue presents the relationship between basic and non-basic rights as simple and straightforward.[15] However, it seems to me that he makes three slightly different arguments for regarding basic rights as necessary for other rights.

I will call Shue's main argument the *constitutive argument*. According to the constitutive argument, every right R includes a putative basic right. For example, the right to assembly includes the right to assemble without being beaten or murdered while doing so. Respecting the right of assembly necessarily involves respecting the security rights of those who are assembled because the former right is partly constituted by the latter rights.[16] The constitutive argument seems to be true for security rights: for any right R, beating or murdering someone trying to exercise R is a way of violating R.

The case appears to be different where subsistence rights are concerned. On the face of it, while assaulting a group of people is a way of interfering with their right of assembly, failing to provide them with health care is not. The right of assembly is the right to gather as a group, it is not the right to gather as a group in good health and the duties corresponding to this right involve protecting groups that seek to gather but do not extend to enabling them to do so. In his discussion of subsistence rights, Shue replaces the constitutive argument with what I will call the *use argument*. According to the use argument, the protection of subsistence rights is necessary for any other right R because people can only use R if their subsistence rights have been met. A person who lacks the means of subsistence can be harmed just as severely as one who is assaulted and the "resulting damage or death can at least as decisively prevent the enjoyment of any right as can the effects of security violations."[17] Shue does not distinguish between the use and constitutive arguments as I have done and I take it that he sees them as fundamentally the same argument. One possible explanation is that he sees the constitutive argument as underwritten by the use argument. One might think that the constitutive argument for security rights makes sense because physical assaults disrupt a person's ability to use rights like the right of assembly. But if that is what the constitutive argument maintains then other impediments to one's ability to use one's rights should count as well. Thus a right's fulfillment may require physical health as well as physical security since, to return to the right of assembly, people cannot assemble if they are too ill or otherwise weakened by poverty to leave their homes. If the constitutive argument is to be understood in terms of the use argument, the case for security rights is, ultimately, the same as the case for subsistence rights.

I will call the third argument that I find in Shue's presentation the *coercion argument*. According to the coercion argument, basic rights are necessary for any other right R because they protect people from being coerced into waiving or failing to exercise R. For example, the right of assembly is not fully protected if people can be threatened with physical violence for legitimately assembling.[18] Similarly, an employer might threaten to bar those who attend a union meeting from the company town's only clinic. In either case, the threat to a fundamental interest in physical integrity or basic subsistence can lead people not to use their other rights. Conversely, protection of basic rights would diminish the threat: it is more difficult for private parties to threaten to deprive someone of security and health care if they are socially guaranteed than it is if they are not.[19]

Do these arguments show that security and subsistence rights are, strictly speaking, basic rights, that is, rights that are necessary for the enjoyment of all other rights? They show that protection against threats to security and subsistence is implicit in many rights. However, they do not show that this protection extends beyond the protection needed to exercise those rights. Consequently, they do not show that everyone who accepts the existence of some right R is committed to accepting rights that offer protection against threats to security and subsistence in general, that is, apart from the protection needed to exercise R, the right whose existence has been conceded.

For example, the constitutive argument's claim about the relationship between rights and security is true but it does not support a general right to security. By a general right to security, I mean a right against physical assaults in all circumstances. Showing that I have general security rights means showing that I have rights against being assaulted while I eat my breakfast, read the paper, walk in the park, type at the computer, or attend an anti-government rally. But the constitutive argument only shows that a right R entails security rights during the exercise of R. If R is the right of assembly, the constitutive argument shows that I have a right to security while I am assembled with others, but not that I have rights to security while alone. Accordingly, it does not show that a general right to security is entailed by any other right.

The use argument also seems to fall short of showing that the full range of subsistence rights are necessary for any other right. This is so because there is a point to many of the civil and political rights even if subsistence rights are not protected; the former are not useless in the absence of the latter. For example, one can certainly enjoy or use a right against torture even if one is in very poor health. Since the difference between being very sick and being both very sick and subject to torture is quite significant, having a right against being tortured will still mean quite a lot to me even if I do not have a right to health care. The same may be said about being subject to punishment for what one says or believes: there is a difference between having rights against these things and not having them even in the absence of security or subsistence rights. I may still have use for a recognized right against physical abuse in retaliation for things that I say or believe even if I do not have the right to physical security in general and, in fact, even if I have no recognized right against being killed. This is so because having protection against one kind of physical abuse is significantly better than having none at all.[20]

Finally, the coercion argument shows, at most, that someone who believes in a right R is committed to believing that there are rights against coercive threats directed at the use of R such as rights against threats to subsistence and security that are tied to the exercise of R. For example, the employer who threatens to block access to the clinic if his employees attend a union meeting would violate such a right. That, however, is much more narrow than a right to subsistence in general. Such a right would not be violated if, for example, the clinic were shut down because of its costs. A general right to subsistence, by contrast, is supposed to guarantee access to health care against both kinds of threats. Nor would a right against threats to my security aimed at forcing me to abandon my other rights show that I have a right against being mugged by someone who only wants my money and is happy to allow me to deliver the speech in my pocket.

For these reasons, the argument for basic rights succeeds only in a highly qualified way. It is not true that the full-blown security and subsistence rights are necessary for the enjoyment of all other rights. Some rights can be enjoyed even without the protection of subsistence rights: this is true of the relationship between the right against being tortured and subsistence rights. Others can be enjoyed with merely partial protection of these rights, namely, protection that is tied to the exercise of the right. My right of assembly only gives me the right to be free from those physical assaults that disrupt my ability to assemble with others. It does not entail a right against physical assault that is unconnected to gathering with others. Consequently, neither subsistence nor security rights are basic rights, defined as rights whose protection is necessary for the protection of all other rights.

However, these limitations may not be fatal for Shue's purposes. Even if it is not the case that any particular right entails general security or subsistence rights on its own, it may still be the case that some of the rights that are acknowledged by those who believe in civil and political rights will entail quite broad security rights. Insofar as a right to liberty or free movement covers a wide range of activities, for example, the constitutive argument would show that there are security rights covering that range of activities as well. It may also be true that the sum of several civil and political rights will work to protect security in a broad range of cases. For example, someone who accepts the standard list of civil and political rights such as the rights to life, personal liberty, liberty of movement, privacy, thought, conscience, expression, assembly, and association would be committed to a very broad right to security according to the constitutive argument. So while the constitutive argument may not show that security rights are basic rights as Shue defines them, he might still succeed in his more fundamental goal of showing that security rights are entailed by other, widely accepted, rights.

Similarly, the assertion that basic rights are necessary for the use of any other right may not be essential for Shue's case. He might relax the assertion that basic rights are necessary for the use of any other right and settle instead for pointing out that they are necessary for the use of some very important or widely accepted rights. For example, while there are some rights, such as the right against being tortured, that can be of use to a person who never leaves his bed, others protect the liberty to lead an active, autonomous life. These rights will be useless to those who are too hungry or ill to enter the world, engage with others, and use their liberty in active pursuit of their own aims.[21] A more modestly formulated use argument would, perhaps, be as effective as the bolder version, provided most of those Shue hopes to address accept suitably broad liberty rights and attach sufficient importance to preserving the ability to lead an active and autonomous life. For the purposes of undermining the alleged priority of civil and political rights, it is not necessary to show that any possible right entails subsistence rights but only that some of the accepted civil and political rights do.

Before closing this section, I should note an important limitation to such an argument. It will not show that those who cannot lead an active and autonomous life have a right to health care. There are some people who are so ill or incapacitated that they are unable to make use of their rights to assembly, speech, or liberty of movement even if they receive adequate care. No matter what is done for them, these people may never leave their beds and engage the world in an active way. An argument for a right to health care that is based on the role of health care in enabling people to use the rights that protect living an active life will not apply to them since they will not lead such lives in any event. This may seem perverse as these are the people who, presumably, are among those with the greatest needs for health care. That suggests that a direct case for a right to health care would be, at least in this respect, more satisfying than the indirect one. If one could argue for the importance of a right to health care on its own, as opposed to showing that it is important because of its relationship to other rights, then one could offer a more persuasive analysis of these sorts of cases. Of course, Shue's indirect argument is not incompatible with a more direct account but, if my analysis of it is correct, it will not have the same implications.

DUTIES

Shue's second thesis is that basic rights have at least three corresponding kinds of duties: deprivation duties, duties not to deprive others of the substance of their basic rights; protection duties, duties to protect others against deprivation; and aid duties, duties to aid those who have been deprived. What is the relationship between the first thesis, about the connection among rights, and the second? That is, can one show that some right R entails all three of the duties that correspond to security and subsistence rights?

At least some of the standard civil and political rights entail duties against certain ways of depriving others of subsistence and security; in that sense, they entail subsistence and security rights. For example, the constitutive argument from the right of assembly to security rights seems to show that the right of assembly entails duties not to assault people who are legitimately assembled since such assaults are a way of violating the right of assembly.[22] Something similar may be true of subsistence rights: depriving people of the means of subsistence can be a way of preventing them from using their right of assembly. For example, I might prevent you from joining an assembly by hiding the medication you need to have the physical strength to march or speak. On a larger scale, I might block the delivery of drugs to kill the parasitic infections that have left the townspeople too lethargic to organize against my boss, the corrupt mayor. These examples are unusual and, no doubt, extremely rare. Nonetheless, they seem to be examples of violating the right of assembly and, therefore, examples of a connection between the right of assembly and a right to health care.

What these examples seem to illustrate is that violating the deprivation duties corresponding to security and subsistence rights can be a way of violating other rights such as the right to assembly. Suppose we grant that this is so and hold that those who accept that there is a right of assembly are committed to thinking that it imposes duties on others not to prevent assemblies either through physical assault or by blocking access to certain kinds of health care. Would we have also shown that those who accept the right of assembly are committed to the other duties that are said to correspond with security and subsistence rights: protection and aid duties? On the face of it, that would not follow. Most people agree that the right of assembly imposes duties against assaulting those who are legitimately assembled. Few people, by contrast, think that the right of assembly entails duties to protect others from being assaulted. We do not seek out assemblies to protect and if an assembly is attacked we do not think that we have violated the rights of the victims simply by virtue of having failed to protect them. Accepting the right of assembly may commit one to duties not to deprive others of their security rights, but it does not obviously commit one to accepting duties of protection or aid corresponding to security rights.[23] Similarly, agreeing that the right of assembly entails duties against withholding health care in order to prevent assemblies does not commit one to accepting duties to protect access to health care or to provide it to those who cannot obtain it on their own.

Note that it is conceded in each case that security and subsistence rights are both entailed by another right. Security and subsistence rights are both genuine rights and have as much priority over other goals as the right from which they are derived. What the examples show, however, is that this is a less interesting result than we might have expected. What appears to be important is not whether we can show that there are rights to security and subsistence but what duties can be shown to correspond to those rights. The interesting question, in other words, is not whether there are basic rights that take priority over other rights or goals but what duties there are and what priority they have relative to other goals. The first thesis, that there is a necessary connection among rights, leaves this issue open.

Perhaps the question of duties only seems to be left open because we have not given due weight to Shue's definition of rights. This definition holds that a moral right "provides (1) the rational basis for a justified demand (2) that the actual enjoyment of a substance be (3) socially guaranteed against standard threats."[24] There are two features of this definition that cut against the objection I have been considering. First, the function of rights, on this definition, is to protect especially important interests, namely, the interests in having substances such as physical integrity, food, health care, and so on. If that is what I accept in accepting rights to subsistence or security, then there is no obvious reason for drawing the lines at any particular kind of duty: any acts or omissions that would protect the interest may be required and there is nothing obviously special about deprivation duties as opposed to the others. Of course, one may object that requiring me to protect or provide for others is generally more burdensome than requiring me to abstain from depriving them of substances. But this is not always the case and the objection will bear little weight if providing for someone else can, say, save a life at a trivial cost. Second, the definition attributes responsibility for protecting rights to societies while the objection to moving from one kind of duty to another involves examples featuring individuals. Perhaps the objection to moving from one kind of duty to another does not apply when the subject is social guarantees. I will take up each point in turn.

Shue's definition of rights does not leave much room for insisting that there is only one kind of corresponding duty. But the definition is stipulated; Shue explains that he means to support it by offering a persuasive account of basic rights but, of course, that account may well be in dispute.[25] Thus, an opponent may avoid being forced to move from political and civil rights to security and subsistence rights with all of their attendant correlative duties by defending an alternative definition of rights. If, for example, a right is identified by the claims one can make against others rather than by its substance, there will be no obvious reason why one would be committed to any duties beyond those acknowledged with the original right.

Even if the definition of rights is granted, however, one may still draw distinctions among the priorities to be assigned to the duties corresponding to security and subsistence rights. The importance of the interests protected by these rights does not automatically transfer to the duties corresponding to them because not all duties have the same bearing on the substance of one's rights.[26] For example, if I fail to do my duty by assaulting you, I directly deprive you of the substance of your rights: you would have lost your physical security. By contrast, failing to perform my duties to provide for others is not incompatible with their continuing to enjoy the underlying substance of their rights. My failure to provide you with health care will not necessarily mean that you will lack access to health care or suffer from ill health: someone else may provide care for you or you may avoid illness on your own. Given the definition of rights, I would violate your rights in either case. In both examples, I fail to perform the duties that correspond with your rights and thus I violate your rights. But the consequences of these failures are different. In the first case, the consequence of my violating my duty is that you certainly suffer harm. In the second case, by contrast, there is only a probability of harm. Given that difference, the duty in the first case is, presumably, more important than the duty in the second case.

It should be added, however, that the different weights attributed to different duties do not line up with the distinctions that Shue criticizes, namely, the distinction between security and subsistence rights and the distinction between negative and positive rights.

For example, I may directly deprive you of food, medicine or water by locking you in a room or hiding these things from you; in that case, violating my deprivation duties corresponding to your subsistence rights would directly harm you. On the other hand, a would-be assailant may lack the physical strength to deprive his victim of her security; he may both fail to do his duty and to deprive his victim of the substance of her rights. Nor is it the case that deprivation duties will always have greater weight than protection and aid duties. The feeble assailant violates his deprivation duties with little consequence. A company that fires its workers may take away their means of subsistence, contrary to its deprivation duties, but they may find work elsewhere. More controversially, one may think that the violation of deprivation duties could have little consequence if a harm is overdetermined. A central bank may restrict the money supply to cause a recession, throwing some people into poverty, without having done a serious wrong: if their plight was inevitable and would have been caused by the economic crash that the bank was trying to head off, the bank's causing them to fall into poverty, arguably, does not make them significantly worse off than they would otherwise have been. By contrast, it is a serious wrong to casually watch an injured child bleed to death while sitting on a crate of bandages even though this is a violation of a duty of aid as opposed to a deprivation duty. Thus an apparently positive protection duty may be more important than an apparently negative deprivation duty.

The upshot of this is that establishing the priority of security and subsistence rights is not the same thing as establishing the priority of the corresponding duties. It may be the case that a duty corresponding to one of these rights is sufficiently unimportant to be outweighed by a competing right, duty, or even consideration of the general welfare. However, this might be an acceptable result for Shue. It does not mean that only security rights are genuine rights or that only security rights have genuine corresponding duties. Nor does it mean that subsistence rights merely have the same priority as desirable, but not mandatory, social goals. Finally, it does not mean that positive duties are always less important than negative duties. A duty to protect someone from immediate deprivation, where there is no available alternative for the victim, could well have similar or greater weight than a duty forbidding direct deprivation.

A second noteworthy feature of Shue's definition of rights is that it concerns social guarantees. The objections raised earlier against moving from duties not to assault others to duties to protect them were based on individual examples. I said that most of us agree that we have duties not to assault others but that few of us think we are required to offer protection. But, on Shue's definition of rights, rights are not claimed against individuals, they are claimed against societies. Does this make an important difference? I think that it might and, in the rest of this section, I will explain why.

One diagnosis of the resistance to moving from deprivation duties to duties of protection and aid is that the first kind of duties, in the cases I was considering, do not have to be allocated to others in the way that the other kinds of duties do. This is a general problem for duties of protection and aid: they cannot be plausibly attributed to everyone in the way that most deprivation duties are but rather have to be allocated to some people specifically. The reason why this is so is that it is hard to see how to formulate these duties so that they are both plausible and fall on everyone. Does every person have a duty to provide for everyone in need? That seems to impose implausibly large demands on resources and knowledge. No one has the resources to provide for all the needy in the world or even all of the needy who may live on a city block. Even if demands on resources were limited, it is hard to see how responsibility could reasonably be attributed: how could I know if I am violating the rights of people far away who may, or may not, have fallen into poverty or lost access to health care and, if I cannot reasonably be expected to know, how can I be held responsible for protecting these people?[27] Conversely, there would be a problem of redundancy: everyone in the world would have the same duty to give aid to a particular person and, if they were actually to follow through, the recipient would be inundated with aid.[28] We may use the circumstances in which a duty bearer finds herself to do the work of allocating duties: the duties fall on those who are able to meet them who are near, in some sense, to those in need. This may be an appropriate way for allocating responsibility in emergencies, but it is, obviously, inadequate for genuinely protecting subsistence and security.

I said that this is a problem for most duties of protection and aid. Many of the central cases of deprivation duties do not face this problem: my right to assembly imposes equal duties on everyone not to beat me while I am assembling with others. But Shue persuasively argues that some deprivation duties may also face allocation problems. Some people may lose access to food and health care as a result of the complex interaction of decisions, policies, and fortune and, in these cases, it is not clear whether any particular actor in such a complex chain of events can be held responsible for the resulting deprivation.[29] There is no obvious party to hold responsible for preventing deprivation in these cases and thus they are unlike the relatively simple case in which we are enjoined not to assault those who are peacefully assembled. Instead, some allocation of responsibility has to be made either to single out responsible parties in the chain of events or to ensure that there is some way of enabling those who have been deprived to regain access to the means of subsistence.

Addressing rights claims to societies rather than individuals avoids some of these problems in the allocation of duties. Societies often have the institutional means either to meet very complicated claims directly or to allocate responsibility for them to particular parties. Social institutions can cut through the problems of excessive demands, redundancy, and knowledge by assigning responsibility for particular cases to particular agents. Of course, social institutions cannot magically expand the resources or epistemic powers to be brought to bear on a problem. But they can enable us to use the resources and knowledge that we have more efficiently. For example, an institution can attribute responsibility for particular problems to particular agents, thus reducing the scope of the problem those agents must address. By the same token, the redundancy problem can be reduced if responsibility is dispersed by means of institutional decisions: if we face a situation in which we know we must all do something but we want to ensure we do not do the same thing, the natural thing to do is to communicate or accept some rule ensuring that we do separate things. Finally, institutions may allocate duties in the case of failures: for example, police forces are responsible for protecting people against those who would assault them. Similarly, if the party who is primarily responsible for providing health care fails, a second party may gain responsibility for doing so. Thus the existence of social institutions may make possible duties that could not be justified in a world composed solely of individuals.[30]

I have covered several distinct points in this section and so it may be worth restating them. First, it seems relatively easy to show that there are at least some security and subsistence rights that are entailed by other rights. What is more difficult is showing that all three kinds of duties corresponding with these rights are entailed by any acknowledged right. Even if this is shown, it is not obvious that any particular duty corresponding with a particular right has the highest priority. Provided one grants the definition of rights that Shue employs, these difficulties do not strike me as fatal for Shue's project: when, for example, it is clear that the failure to provide health care will result in death or serious harm, the duty to provide the care can be as important as a duty not to harm someone through physical assault. Finally, Shue's definition of rights implicitly includes a way of addressing an important problem with allocating duties of protection and aid.

FAIRNESS

In the previous section, I argued that most duties of protection and aid have to be allocated in a way that most deprivation duties do not. It is natural to turn to societies or governments as the entities that ultimately bear responsibility in these cases on the grounds that they have the institutional means to address these problems on their own or to assign responsibility for them to others. However, while attributing responsibility to a society is a way of reducing some of the problems with attributing duties directly to individuals, it is not fully satisfying on its own, without further elaboration. First, attributing responsibility to societies is fine as far as it goes, but there are clearly better and worse ways of allocating responsibility and, ideally, one would have something to say about how societies should do so. Second, there are many cases in which responsibility cannot stop at the social level, provided the aim is to guarantee protection of certain interests. For example, a state may be the main threat to security, a society may not be organized enough to carry out an effective public health program, and, even if it is, it may not own the drugs or vaccines it needs to address pressing health problems. In some of these cases, the proper allocation of duties may be unclear: for example, should drug companies be forced to share important drugs and vaccines with poor countries at a loss, should wealthy countries pay them to do so, or should some other scheme be devised? There is no obvious allocation of responsibility here and the institutional or conventional solutions that seem to be called for are beyond the powers of the societies whose members need them the most. More generally, insofar as security and subsistence are thought of as human rights, it is hard to see why they would be the exclusive responsibility of societies. On the face of it, they are the responsibility of humanity as a whole and societies should be held as the primary bearers of responsibility only as a matter of convenience or because there is no better alternative.

Of course, there is no global institution analogous to a state that might bear the responsibility for settling questions about the allocation of duties. Given that, one might ask whether basic rights should really count as human rights. So long as there is no global allocation of duties, one might say, there is no way to issue claims based on basic rights against the world since there is no particular agent that bears the duties to respond to the relevant claims. But the claim that an institutionally defined allocation of duties is a necessary condition of making claims at all strikes me as too strong. The usual situation is that there are some individuals or societies who are capable of addressing the needs of those who lack adequate security or the means of subsistence. The latter can address a claim to the former. What is often at issue is whether those claims can be resisted on the grounds that there is nothing special about the individual or society being addressed such that he or it, rather than another individual or society, must respond to the claim or on the grounds that it would be unfair to expect the individual or society to respond when others will not do their shares. Both problems are especially prominent in the absence of institutions that can allocate responsibilities to particular agents and ensure that all agents do their share.

Shue takes up concerns about the allocation of duties in the postscript to the second edition of *Basic Rights*. In particular, he suggests that a necessary condition of there being such duties is that they are allocated fairly even though he worries that a fair allocation of duties might not provide optimal protection for basic rights.[31] The specific sort of problem that worries him concerns the devolution of duties. If A has a duty to protect B's basic rights and fails to perform it, does C then inherit A's duties? B's rights are still unmet, after all. Holding C responsible seems both unfair and unmotivated since A is responsible for B's plight, not C.[32] Furthermore, if duties automatically pass down in this way, they would all quickly fall on those who are especially dedicated to human rights and that seems to impose an unfair burden on them.[33] As a response to this kind of problem, he considers a cutoff: when some duty bearers do not do their parts, others gain additional duties to protect the rights holders from violations of their rights but not necessarily additional duties to provide aid to those whose rights are violated.[34]

I do not doubt that fairness is an important concern. Other things equal, it is better that a distribution of duties be fair than unfair. But should fairness be a necessary condition of duties corresponding to basic rights? This condition stands in tension with the characterization of basic rights as rights to the most rudimentary things that people need to live decent lives: physical safety, food, health care, and shelter. Shue aptly describes basic rights as "the moral minimum ... the least that every person can demand," "the line beneath which no one is allowed to sink," and "everyone's minimum reasonable demands upon the rest of humanity ... the denial of which no self-respecting person can reasonably be expected to accept."[35] If that is really so, and it is hard to believe that it is not, why should the unfairness of bearing duties corresponding to these rights be a sufficient reason to resist those duties? By hypothesis, B is demanding the least that can be demanded of C. C may say that B really should address A, since it is A's responsibility to respond to B. But if A simply will not do his part, B has no other alternative, and C has no other compelling reason to resist the duty, why should the unfairness of putting the duty on C outweigh B's very minimal demand? B may agree that C has a complaint, but, it is against A, not B since it is A's failure to do his duty that is the reason why C is called on to help B. Given that B has done nothing wrong to C, why should B be allowed to sink below the very minimal line drawn by basic rights simply because of C's complaint against A?

This is a significant issue. A fair allocation of duties corresponding to basic rights will, presumably, depend on institutional arrangements that will allocate specific duties to specific agents. In the absence of institutional arrangements, it will be extremely difficult for anyone to know whether he has done his fair share, or even what his fair share involves, and it will be highly likely that some will not do what they regard as their fair share. But such institutional arrangements, if they are ever made at all, will be a long time in coming. While we wait for the relevant institutions to be developed or devote our energy to building them, basic rights will go unmet. Given the significance of basic rights, it is not at all clear that achieving fairness is worth the price. Thus it is not clear that individuals and societies can resist duties corresponding to these rights on the grounds that they are unfairly allocated. Nor is it clear that resources that might directly protect the relevant interests should be invested in seeking institutional solutions to the problem of allocating and enforcing duties in a fair way.

I do not maintain that this is a serious problem with Shue's fundamental theory. Fairness is a genuine concern and the proposal is, as he puts it, "completely tentative."[36] My reason for bringing the matter up is that the tension between basic rights and fairness is interesting because it suggests that justice and rights can come apart. If B's claim against C were based on justice, that is, if B were saying that C would be unjust to refuse his demands, then his position would be a weak one. Justice requires that each do only his part or what is fair. But C, by hypothesis, has done his fair share. So while B's complaint of unjust treatment could be directed at A, it could not be directed at C. Nonetheless I concluded that B has a claim against C based on his basic rights: even though C has done his fair share, it would be indecent to refuse B's request for help. It follows that demands based on basic rights are not demands based on justice. In fact, one can make unjust demands on the basis of one's basic rights in the sense that one can demand that people do more than justice requires by bearing unfair burdens.

I began with an assumption that one way of deciding whether inequalities in health care are unjust is to discover whether there is a human right to health care. That assumption about the relationship between justice and rights stands in need of qualification. The assumption is not false, because it may be the case that unequal protection of basic rights to health care is unjust and that those who have failed in their duties have been unjust towards those who lack adequate health care. The qualification that seems appropriate is that it may also be the case that a claim based on basic rights could run contrary to what justice requires. Since that is so, while there may be a connection between justice and rights, the two are not identical and may even come apart. Thus we may conclude that inequalities in health care are unjust but that the only realistic ways of redressing this kind of injustice go beyond what justice requires or even that they involve imposing unfair burdens on some.

NOTES

[1] I do not mean to contend, as Mill seems to, that justice includes all rights with correlative duties. See John Stuart Mill, *Utilitarianism*, Edited, with an introduction by George Sher. (1861; Indianapolis: Hackett Publishing Company, 1979) 48-9; for criticism, see David Lyons, *Rights, Welfare, and Mill's Moral Theory* (New York: Oxford University Press, 1994) 143-46. I assume that the case for thinking that the unequal protection of human rights is a matter of injustice is stronger than the case for thinking that justice concerns all rights.

[2] Henry Shue, *Basic Rights: Subsistence, Affluence, and U.S. Foreign Policy*, Second ed. (Princeton: Princeton University Press, 1986) 13.

[3] Ibid. 15.

[4] Ibid. 32-33.

[5] Ibid. 16-17.

[6] Ibid. 19.

[7] Ibid. 31.

[8] Ibid. 6.

[9] These are some of the rights listed in the International Covenant on Civil and Political Rights. Social and economic rights, as given in the International Covenant on Economic, Social, and Cultural Rights, include rights to work, decent working conditions and pay, trade unions, social security, food, clothing, housing, health, education, and participation in cultural life.

[10] Shue, *Basic Rights* 52-55.

[11] Ibid. 156; Jeremy Waldron, *Liberal Rights: Collected Papers, 1981-1991*, (Cambridge, England; New York: Cambridge University Press, 1993), pp. 25, 211-15.

[12] Shue, *Basic Rights* 20.

[13] Ibid. 23.

[14] Ibid. 9, 22-23, 24, 25, 25-26.

[15] Ibid. 20-22.

[16] See, for example, Ibid. 27.

[17] Shue, *Basic Rights* 24.

[18] See, for example, Ibid. 26.

[19] It would also have to be the case that these rights could not be forfeited or waived by those who hold them. Shue does not discuss this sort of issue in the main text, but he does seem to commit himself to the opposite position in a footnote. See Ibid. 186-87.

[20] Shue credits Mark Wicclair with raising a similar objection; his response to the objection appeals to the coercion argument. See Ibid. 184-5.

[21] Waldron appeals to this sort of case in support of Shue's position. See Jeremy Waldron, "Liberal Rights: Two Sides of the Coin," in *Liberal Rights* (Cambridge: Cambridge University Press, 1993), 7-8.

[22] I will continue referring to the right of assembly in what follows. While I do not think that any one right could entail general rights to security and subsistence, I assume that any connection between this right and security or subsistence rights can be generalized to broader rights of liberty and movement along the lines described at the end of the last section.

[23] Compare Shue, *Basic Rights* 43.

[24] Ibid. 13.

[25] Ibid. 183.

[26] Compare Ibid. 111-19. See also Jeremy Waldron, "Rights in Conflict," in *Liberal Rights* (Cambridge: Cambridge University Press, 1993), 215-20.

[27] Compare Samuel Scheffler, "Natural Rights, Equality, and the Minimal State," in *Reading Nozick*, ed. Jeffrey Paul (Totowa, NJ: Rowman and Littlefield, 1981), 162.

[28] For the resource and redundancy objections, see Carl Wellman, *Welfare Rights* (Totowa, N.J.: Rowman and Littlefield, 1982) 159-63.

[29] Shue, *Basic Rights* 44-5, 58-9.

[30] For a similar argument, see James Griffin, "Welfare Rights," *The Journal of Ethics* 4 (2000): 31-32.

[31] Shue, *Basic Rights* 164-66.

[32] For discussion of this kind of problem, see: L. Jonathan Cohen, "Who is Starving Whom?," *Theoria* 5, no. 2 (1981): 65-81; Liam B. Murphy, *Moral Demands in Nonideal Theory*, Oxford Ethics Series (New York: Oxford University Press, 2000); and James Rachels, "Killing and Starving to Death," *Philosophy* 54 (1979): 159-71.

[33] It would also give perverse incentives to those who are less committed to human rights to fail in their duties, knowing that others will take care of the resulting harms. Shue, *Basic Rights* 172-73.

[34] Ibid. 173.

[35] Ibid. xi, 18, 19.

[36] Ibid. 173.

ROSEMARY B. QUIGLEY

ADVOCACY AND COMMUNITY: CONFLICTS OF INTEREST IN PUBLIC HEALTH RESEARCH

ABSTRACT: The players in modern biomedical and public health research are entwined in new and complicated ways. As industry established cooperative research efforts with clinicians and other scientific investigators, the financial influence of these arrangements on the ethical conduct of research has received increased scrutiny. Though many have regarded the cooperation of patient advocacy groups and community-based organizations as a mode of protecting research participant interests, as the influence of financial interests spreads these entities may also become suspect in pressing an agenda on behalf of those they purport to represent. This essay describes the way these influences may take hold to influence intermediaries between investigators and participants in biomedical and public health research settings. Researchers must recognize their obligation to inform and empower research participants directly; the involvement of advocacy or community-based groups cannot supplant this responsibility.

KEY WORDS: Research ethics, advocacy, public health research, informed consent, community-based research

INTRODUCTION

Advocates have long been viewed as protectors of more vulnerable populations in the scientific quest for information about the health and welfare of individuals. They are presumed to represent the patient or community member in many respects, including personal perspective and socioeconomic circumstance. It is the advocate who generally pushes a research agenda, bringing attention to their cause and seeking the evidence to support the reforms that are sought. Indeed, the advocate for a particular group has come to entirely supplant members of the group itself, as advocates are the more willing, dedicated and vocal counterparts to the group members and their interests. In this sense, researchers seeking access to study populations more readily involve advocates than lay community members, who may be difficult to reach, educate and incorporate into discussions about the design and initiation of research protocols.

The legitimacy of the advocate, who is often replete with noble intentions, is rarely questioned. But the truth is that the advocates may have interests distinct from the community they purport to represent. They may seek the spotlight for themselves

223

M. Boylan (ed.), Public Health Policy and Ethics, 223-235.
© *2004 Kluwer Academic Publishers. Printed in the Netherlands.*

as much as their issue. And many advocacy groups, or even community-based organizations, face the same economic pressures of other businesses. In this respect, their priority-setting may be skewed in ways that do not clearly reflect the interests of individuals or populations. This essay explores the tradition by which advocates have come to represent others' interests and explores how the emerging concerns surrounding conflicts of interest in biomedical research may have overlooked the crucial role of advocacy in these dynamics. The discussion goes on to suggest the ways in which those who style themselves as representatives of community interests in the facilitation of public health research may actually malign vital research objectives in terms of setting research agendas and protecting research participants.

BIOMEDICAL RESEARCH AND THE ROLE OF ADVOCACY

The emergence of advocates as a force in shaping biomedical research agendas has marked a pivotal point in how responsive research is to public demands. To some extent, advocates did not even exist forty years ago, as patients still deferred to physicians about the best course of action in medical treatment. But as autonomy emerged as a central principle in the delivery of patient care, vocal groups banded together to press for better treatments and improved chance for survival via medical advance. Advocates have come a long way to be included in assessment of research priorities and oversight of human research protections. Still, those selected to participate on institutional review boards (IRBs) overseeing the ethics of clinical research or committees informing the research agenda are usually known quantities, players in the advocacy community who, while representing a patient or family voice, may still not represent the mainstream experience of the patient community.[1]

The classic conception of the advocate in the biomedical context has several components. First, an advocate seeks to raise awareness and understanding about a specific condition among the public. In addition, the advocate is poised to lobby on behalf of those with the condition, whether on health policy and insurance issues or medical care and research interventions. Advocates also raise research monies to investigate the causes of particular conditions and the potential for therapeutic improvements. Advocates will usually represent a group of patients with a common diagnosis, but advocacy organizations have increasingly joined consortia with other diseases and conditions to increase the scope of their influence. Also, advocates have recently been tasked to individual patients rather than patient groups, for instance supporting a patient through a research protocol or a particularly challenging medical intervention.

This model advocate emerged in the 1980s, as communities afflicted by AIDS bore outspoken individuals who increased the profile of the AIDS research agenda within the scientific establishment and pressed for the widespread enrollment of patients into clinical trials as a means to accessing the best, cutting-edge therapies.[2] The development of strong advocacy groups for breast cancer patients and survivors arguably transformed the methods of treatment available to this population, with research into less aggressive surgical interventions in combination with chemotherapy and radiation evidencing comparable or improved life expectancy.[3] Advocacy groups are cobbled together from various individuals with investment in a given condition, sometimes those suffering the condition, but more often those who care about those suffering the condition, such as parents, spouses and children of a present or past patient.

While these forerunners' efforts were directed at federal funding agencies, such as the National Institutes of Health, and some academic researchers, this focus has shifted somewhat in the past decade. Increasingly, medical advocacy organizations are lobbying just as hard within industry for researchers' attention to their disease process of interest. The burgeoning portion of the biomedical research budget supported by industry has necessitated this redirection. Advocacy organizations are increasingly using funds raised as seed money for private biotechnology companies to pursue novel research concepts. For instance, the Cystic Fibrosis Foundation has established a research arm that grants money directly to companies and researchers, with oversight of the research conducted by the Foundation.[4] Likewise, in the alpha-1 antitrypsin deficiency community the advocacy organizations have splintered, with the more recently established Alpha-1 Foundation focused on providing monetary support for research, creating patient registries for research, and sponsoring research symposia.[5]

To some extent, the evolution in the priorities of advocacy organizations has created a schism between the advocates and the patient populations they represent. These organizations often fashion their fundraising efforts around the goal of achieving a cure for a particular disease. Somewhat less satisfying is the objective of improving the lives of individuals suffering a given condition short of a cure, for instance by developing mitigating therapies or providing needed social supports. The conflict of priorities pits helping those with the disease at present against curing those who will develop the disease in the future. Another layer of this conflict may be those organizations that advocate prevention of the disease or condition in the first place, such as through genetic screening or behavioral change. Which objective an organization chooses as its first priority certainly reflects a value judgment about the lives of current patients, and patient communities may logically feel forsaken and exploited by their supposedly representative advocacy groups when prevention and cure are increasingly emphasized. As such, the priorities of the advocacy groups may not align substantially with those of individual patients, raising questions about the capacity of these organizations to represent patients in discussions about research agendas.

CONFLICTS OF INTEREST IN BIOMEDICAL RESEARCH

Recent controversy about the protection of human research participants has brought the pressures constituting conflicts of interest in biomedical research into public view. Significantly, most of the attention has been paid to conflicts of investigators pursuant to their relationships with industry. For instance, in the case of Jesse Gelsinger, the young man who died in a gene transfer protocol for OTC deficiency at the University of Pennsylvania, this concern was characterized by the principal investigator's significant equity interest in the private company that would be developing the therapy commercially.[6] In addition to the classic case of an investigator holding equity interest, financial incentives may compromise investigator's objectivity where more positive or significant research results are sought by an industry sponsor, or where investigators depend on a flow of funding from industry to continue their academic research initiatives.

There has also been a surge of attention paid to nonfinancial conflicts of interest.[7] As industry-sponsored research becomes as much a badge of honor in the biomedical research establishment as publicly funded research, the approbation of peers and recognition within academic institutions also becomes an incentive that may spur investigator conflicts. Investigators may also be hamstrung in disseminating research findings when the institutional contracts with a research sponsors have secrecy provisions that give the sponsoring companies discretion over release of study results.[8]

Clinical investigators are often in the position of providing treatment for the same patients they are referring to clinical trials.[9] There are those who contend that this relationship is at the root of patient confusion about the therapeutic nature of clinical research, as patients presume that their personal physician would not recommend a clinical trial unless there was notable potential for benefit, or at least avoidance of harm.[10] The reality is that these clinical investigators truly wear two hats, offering proven therapies but also seeking innovative therapies through the equipoise of research. Patients often do not appreciate this distinction. In fact, they may be flattered by the physician's suggestion that the patient is compliant and altruistic enough to participate in a research regimen. Patients may also sense the possibility of an ancillary benefit to acceding to the request of a physician, such that the physician will be more involved with that patient's case in the future as a form of reciprocity.

The untold story in conflict of interest discussions may be the role of advocacy organizations. Just as physicians and clinical investigators have developed closer relationships with industry, advocacy groups have done the same. Like physicians, advocacy groups are generally viewed by patients as a credible source of information related to treatment options and research progress. When advocacy groups boost the potential for treatment advances resulting from ongoing clinical trials, enthusiasm about those trials rises in the patient community. The reliability of this dynamic is significantly undermined if industry's influence mars the agenda of advocacy organizations. But as relationships with investigators and research institutions come under increased scrutiny, industry representatives may be increasingly interested in finding new routes to the research participants they require to conduct sufficient research to satisfy regulatory requirements. The next route to be exploited is via advocacy organizations.

Contemporary advocacy organizations are often in alliance with clinicians who are prominent in the field. These physicians, in turn, are often cooperating with industry in the development and testing of new clinical interventions. The promotion of research, then, becomes a priority of the advocacy group by virtue of clinicians' need to maintain a research portfolio. It is also not uncommon for advocacy groups to now involve actual industry partners in the direction of the organizations' efforts. Some of these industry players will be seeking new opportunities for investment and fertile ground for conducting research. Last but not least, the organizations themselves may have a substantial interest in fostering alliances with industry, especially when they receive remuneration or other favor for promotion of research protocols within the patient population. As with most not-for-profit organizations, disease advocacy groups struggle to maintain the solvency to afford the overhead of professional staff and patient services. Alliances with industry may be instrumental to the survival of these organizations, and advocates may view this concession as a case of means justifying ends.

In some tight-knit research communities, participating in research is seen as a duty. Those patients who opt not to contribute in this fashion may be excoriated by peer patients, as well as clinicians pushing their research agendas. Advocacy groups put enormous stock in research, promoting it as the key source of hope for current sufferers of conditions. In many cases, however, clinical trials are testing interventions of unproven benefit that may be unlikely to lead to innovative therapy in the lifetime of the patient. Still, the close link between research and hope is exploited in some circles to promote research enrollment.

To the extent that this tactic is adopted by advocacy organizations due to the pressures of industry affiliates, the integrity of these organizations is jeopardized. Patients, then, are in the position of once again being manipulated by an entity that they have every reason to trust, based on traditional sentiments about the role of advocacy. It is not a stretch to imagine that as advocacy groups become funders of industry research, they too will be bound by some contract arrangements that may restrict full disclosure of research findings. Indeed, advocacy groups themselves are not well-served by unsatisfactory research results in the endeavors they support, and may also have some interest in emphasizing positive versus negative outcomes. In this environment, advocacy groups' purported single-minded pursuit of patient interest cannot be relied upon when conflicts of influence arise.

EMERGENT ATTENTION TO PUBLIC HEALTH RESEARCH ETHICS

As with research involving patients, studies performed for the benefit of populations have come under increased scrutiny. The recent case involving the Kennedy Krieger Institute (KKI) in Baltimore highlighted the importance of adequate review of study design, especially where vulnerable populations are being exposed to some level of personal risk. A failure to acknowledge such heightened concerns will only result in a delimiting of research opportunities and increased oversight of protocol machinations, as with the court's purported tightening of restrictions on pediatric research in *Grimes v. Kennedy Krieger*.[11] One component of bolstering the profile of public health ethics is to raise attention to human research protections in more broad-based population research.

While it is seemingly easy to substitute populations for patients in this paradigmatic discussion, the fact is that the goals of medical ethics and public health ethics have been considered divergent.[12] Medical ethics has the autonomy of the individual as its central value, making issues such as research protection particular to individual clinical trial enrollees. Participants' rights to information, understanding and voluntariness are emphasized by the requirements of informed consent for trial enrollment. It should not be such a stretch for the same rights to adhere to those participating in public health research, which may be characterized as more observational epidemiological studies. However, this research is backed by the goal of improving the condition of populations and a lack of focus on individual interests may draw researchers off valuing the informed consent of the research participants.

One of the innovations in public health research has been to establish community-based partnerships as a way to reach the populations in which valuable research may be conducted.[13] Acknowledging and honoring the different agendas of various partners is viewed as essential to these cooperative efforts.[14] It may be the case that involvement of these community representatives in the design, initiation and conduct of research may be viewed by some investigators as sufficient consideration of individual interests. However, like patient advocacy groups, community-based organizations purport to press agendas on behalf of the population in which they are embedded. They lobby legislative and administrative bodies for what they consider positive change, such as health and environmental provisions that may boost the quality of life in a community. Some community-based groups have their own research agendas and they will recruit professional groups ranging from governmental entities to university faculty to for-profit companies to conduct this research.

In many cases, pursuit of a research initiative has the potential to infuse revenue into a community, for instance by grant to the local city budget. In other cases, community-based groups may receive various forms of direct compensation in exchange for support and facilitation of public health research endeavors, ranging from monies to improve social services they engage in or in supports to facilitate the community mission, such as through donations of computers or foods. In any case, attention to a community need from an attentive external group is a rare opportunity that most community leaders would seek to seize. Under these circumstances, researchers can co-opt the stature of the community-based groups in pressing research participation and cooperation, essentially riding established coattails to population trust. Like patient advocacy groups, the sort of community-based organizations that act as community representatives in public health research often hold the complete trust of their local citizenry, especially when intermediaries are religious groups or educational institutions. To the extent that organization representatives like church leaders are providing access to research subjects, purportedly representing their interests, these representatives assume the ethical obligations of the researchers in making sure that, despite other interests, the autonomy of the individual citizen agreeing to participate in research is honored.

In *Grimes*, the investigators seem to have had a noble research objective – promoting lead reduction efforts in urban housing, with an eye to improving the health status of children in these inner-city communities. As the court explained, "The ultimate aim of the research was to find a less than complete level of abatement that would be relatively safe, but economical, so that Baltimore landlords with lower socio-economical rental units would not abandon the units." This endeavor involved comparing the long-term efficacy of comprehensive lead-paint abatement with potentially more cost-effective repair and maintenance interventions for reducing lead in households. Consequently, the research involved keeping some children in household settings where optimal abatement was not achieved and where significant lead levels appeared in the children's blood samples. The research was governed by institutional review, jointly sponsored by the Environmental Protection Agency and the Maryland Department of Housing and Community Development, and supported by even more local entities like the Baltimore City Health Department. None of these offices prospectively identified the objectionable risk to which the child participants would be subjected in this study.

The principle liaison to the community participants was the landlords, perhaps not models of trust, but parties with which citizens are nonetheless compelled to deal. As compensation for the properties to be used and in return for targeting families with young children as tenants, KKI assisted the landlords in applying for and receiving grants and loans to perform the levels of abatement that KKI needed to study each class of home. Among the low-income families, parents were enticed to enrolling their children with food stamps, money and other items.[15] It is not clear whether there was sufficient effort on the part of the researchers to involve the community in the development and execution of the research, but on the individual research participant level the court's findings demonstrate that the informed consent process did not adequately clarify the ongoing risks to which some classes of participants would be exposed. Some commentators cited the shortcomings of the KKI researchers in failing to form adequate community partnerships.[16] But depending on who the investigators had pursued for support, the community representatives may have been so eager to have someone attend to the lead problem that they would have disregarded the fact that their fellow citizens were subjected to the obscured risks as a means to eventual improvement of the situation.

Policies surrounding public health research would seem to have addressed some of these concerns, for instance with the increased review of epidemiological research by institutional review boards and the inclusion of lay members on such review boards. But just as with other debates about the efficacy of IRBs, the possibility that a sole community representative can raise awareness of population needs within an IRB discussion may be overestimated. And just as discussions surrounding international research have questioned the cultural relativism of imposing Western ethical norms onto foreign or native populations,[17] to some extent application of these norms in community settings may also pose problems. It is the consent process on the ground and the actualized autonomy of research participants, not the sign-off from community representatives, that makes research in a population ethically acceptable.

Putting the results of research to productive use is a vital piece of the promice researxhers make to participants. Under circumstances where there is considerable concern that a disadvantaged population is being disproportionately exposed to research risk, utilizing the results of research to improve the conditions within a community may be a special responsibility inhering in public health research. This is particularly true in cases where improvement is the carrot held out to elicit participation. To some extent, the integrity of the informed consent process is undercut when policy change does not emerge as a benefit of research.[18]

BLURRING OBJECTIVITY AND ADVOCACY IN PUBLIC HEALTH SCIENCE

The mission of public health research may be inherently conflicted in a tug between objectivity and advocacy. Some have even argued that epidemiological research cannot be remotely influenced by an advocacy agenda, otherwise the research's design and conduct become necessarily suspect.[19] But to some extent the very motivators of scientific inquiry bring the neutrality of any investigation into question. "Epidemiology is, like any science, at once objective (using defined, rigorous, and replicable methods to assess refutable propositions) and partisan (reflecting underlying values and assumptions guiding conceptualization, choice, and analysis of research problems)."[20] In another vein, public health research may be construed as troubled for what it fails to do. Much epidemiological work is observational in nature and may involve documenting some persistent harm in a population in order to provide data as a tool in an advocacy agenda. Some researchers may feel stagnant in improving the situation when their research does not involve an intervention that is targeted at improving the conditions of the population.

As with biomedical research, public health research is dependent to a great degree on the agendas of the relevant funding sources,[21] which also raises concern about the potential for certain research findings having any impact. Public health researchers will in some cases be beholden to the requests for applications (RFAs) issued by government agencies for funding. However, there are times when calls for more research may actually be a means to stalling a well-established innovation that is being opposed for reasons distinct from population well-being. As one commentator has observed, "The challenge for advocacy here is to avoid being entrapped by carefully engineered attempts by such opponents to frame debate in interminable 'more research is needed' policy bogs, while at the same time never going beyond the science that underpins sound heath policy."[22]

A recent report from the National Academy of Sciences (NAS) on climate change research provides an example. The NAS panel concluded that it was inappropriate for the Bush Administration to target more monies at researching the global warming problem because the threat is already well-defined.[23] The report suggests that more research funding would only delay necessary governmental action, such as regulation of fuel emissions and big industry, to address the critical environmental problem. Indeed, the environmental health researchers likely to pursue this funding would be committed to the goal of addressing global warming, and would be vying for funding of their research career. However, the power of both goals might obscure the fact that their acting on the RFA may do more harm than good in advancing the environmental agenda. The public health research establishment is no doubt filled with projects reflecting further clarification of largely settled questions. Another recent example was the push for more research on occupational safety and ergonomics standards, which only verified conclusions of the NAS but also lead to delay and withdrawal of federal rules on the issue.[24] To the extent that scientific issues are rehashed as a stalling tactic, the ethical status of this research should be reevaluated, so that resources may be directed at implementing the recommendations derived from valid, existing research results.

While the researcher has a responsibility to select topics for investigation carefully,[25] responding to an RFA seems a fairly benign misstep compared with some of the more brazen deceptions that have occurred under the auspices of advocacy in recent public health debates surrounding tobacco regulation. For instance, the tobacco industry funded a research organization, called the Center for Indoor Air Research. Research performed by this group was either peer-reviewed or "special-reviewed," with the specially reviewed projects constituting the bulk of research on second-hand smoke, which was directed at the industry objective to convolute findings of harm by environmental exposure and defeat efforts to regulate smoking in public places.[26] On another front of the industry's fight to discredit research establishing a link between health risks and passive smoking, companies promulgated a program of "good epidemiology practices" through an advocacy group it funded at its inception, the Advancement for Sound Science Coalition.[27] In an effort to influence professional opinion, educational programs sponsored by the coalition featured passive smoking research as a case study of unsound science. Such collusive efforts by industry permeated the public policy debate about restricting smoking practices. In a review article, researchers found that 74% of those who disclaimed a relationship between health effects and second-hand smoke had ties to the tobacco industry.[28] The regulatory objective sought by public health researchers was ultimately achieved, but perhaps more through social movement than by persuasive scientific findings.

Coercion and harassment of medical and public health researchers is not new, but the situation may be becoming more acute.[29] Special interests may have not only economic investment in advocating a particular position, but emotional, moral and political motives as well. These noneconomic factors have been a main barrier to implementing needle exchange programs, widely lauded as a means to curtailing the spread of disease in drug-using populations.[30] Whatever the motivation, it has become increasingly difficult to parse the agendas of public health advocacy groups from the perception they project. Misnomers applied to advocacy groups, for instance, may assist in fundraising even if the donors do not realize what they are supporting. Some examples would be the Food Chain Coalition, which represents the pesticide industry in opposing regulations,31 and Doctors for Integrity in Research and Public Policy, which opposes gun control and handgun research. Given their veil of social concern, such groups have great power to assail research findings, thus fostering doubt on implementation of corresponding policy recommendations.

Public health investigators have an obligation to monitor the use and manipulation of their research results, with an eye to providing appropriate clarification and defense. Public health research is subject to "the pressure to use science to justify policy (even when the data are inadequate), as well as the vulnerability of science to attacks driven by vested-interests that exploit scientific uncertainty to deflect attention from what is known and the actions that would credibly follow that knowledge."[32] Unfortunately the battle for effective public health policy is about much more than good data. Researchers should thus be aware of the political realities, and steel themselves to endure the throes of debate with rigorous research results in hand.[33]

CONCLUSION

The entanglement of various players in the future of medical and public health research will pose significant questions about who bears what responsibility to those patients and populations who give of themselves in an effort to improve human health. At the very least, researchers must be aware that those who purport to represent the research subjects may not always reflect the interests of individual trial participants, whether for financial or political reasons. Advocacy should not be discounted as a trustworthy and valued player in the development of health policy, as the those pressing for better social conditions have the energy and focus to achieve their goals at least part of the time. But the motives of advocacy groups may sometimes be influenced by increasingly powerful industry pressures, and it may be the role of researchers to monitor such influences and support advocacy groups in concentrating on fulfillment of their noble missions. It will also be necessary for researchers to do better at accessing the actual views and goals of the patient or population in the appropriate conduct of research. This will require additional effort and resources, to bring these individuals into discussions so that their valued perspectives may be elicited.

The standard of informed consent for enrollment in research cannot be glossed over because a community-based group or advocacy organization has said it will be satisfactory to do so. No matter how emerging influences may create conflicts of interest to distract researchers, their principal obligation is to preserving the rights of the research participants. Awareness of where these influences may lie and how they may function will help researchers in fulfilling their role as stewards of public trust even as they effect positive change through enlightening research endeavors.

NOTES

[1] Caroline McNeil, "Cancer Advocacy Evolves as it Gains Seats on Research Panels," *Journal of the National Cancer Institute.* 93 (2001): 257-59.

[2] Rebecca Dresser, *When Science Offers Salvation: Patient Advocacy and Research Ethics* (New York, NY: Oxford University Press, 2001).

[3] Barron Lerner, *Breast Cancer Wars: Hope, Fear, and the Pursuit of a Cure in Twentieth-Century America.* (New York NY: Oxford University Press, 2001).

[4] Cystic Fibrosis Foundation, Therapeutics Development Program, http://www.cff.org/research/therapeutics_development_program.cfm. (Accessed May 29, 2003).

[5] Jon F. Merz, David Magnus, Mildred K. Cho & Arthur L. Caplan, "Protecting Subjects' Interests in Genetics Research," *Am. J. Hum. Genet* 70 (2002): 965-71.

[6] Sheryl G. Stolberg, "The Biotech Death of Jesse Gelsinger." *N.Y. Times Mag.*, (Nov. 28, 1999): 136.

[7] Norman G. Levinsky, "Nonfinancial Conflicts of Interest in Research," *New Eng. J. Med.* 347 (2002): 759-61.

[8] Kevin A. Schulman, Damon M. Seils, Justin W. Timbie, Jeremy Sugarman, Lauren A. Dame, et al, "A National Survey of Provisions in Clinical-Trial Agreements between Medical Schools and Industry Sponsors." 347 *N. Engl. J. Med.* 347 (2002): 1335-41.

[9] Robert Mannel, Joan Walker, Natalie Gould, Dennis Scribner, Scott Kamelle, et al. "Impact of Individual Physicians on Enrollment of Patients into Clinical Trials." *Amer. J. Clin. Oncology* 26 (2003): 171-73.

[10] Rebecca Dresser, "The Ubiquity and Utility of Therapeutic Misconception." *Soc. Philo. & Policy* 19 (2002): 271-94.

[11] *Grimes v. Kennedy Krieger Inst., Inc.,* 782 A.2d 807 (Md. App. 2001).

[12] Nancy Kass, "An Ethics Framework for Public Health." *Amer. J. Public Health* 91 (2001): 1776-82; and Daniel Callahan & Bruce Jennings, "Ethics & Public Health: Forging a Strong Relationship." *Amer. J. Public Health* 92 (2002): 169-76.

[13] Institute of Medicine, *Healthy Communities: New Partnerships for the Future of Public Health.* (Washington, DC: Institute of Medicine, 1996).

[14] Baker EA, Homan S, Schonhoff R, Kreuter M, "Principles of Practice for Academic/ Practice/Community Research Partnerships" *Amer. J. Prev. Med.* 16 (1999): 86-93.

[15] Leonard Glantz, "Nontherapeutic Research with Children: *Grimes v. Kennedy Krieger Institute.*" *Amer. J. Public Health* 92 (2002): 1070-73.

[16] Anna Mastroianni & Jeffrey Kahn, "Risk and Responsibility: *Grimes v. Kennedy Krieger*, and Public Health Research Involving Children." *Amer. J. Public Health* 92 (2002):1073-76.

[17] Leslie London, "Ethical Oversight of Public Health Research: Can Rules and IRBs Make a Difference in Developing Countries?" *Amer. J. Public Health* 92 (2002): 1079-84.

[18] Kass, *op. cit.*

[19] David Savitz, Charles Poole & William Miller, "Reassessing the Role of Epidemiology in Public Health." *Amer. J. Public Health* 89 (1999): 1158-61.

[20] Nancy Krieger, "Questioning Epidemiology: Objectivity, Advocacy, and Socially Responsible Science." *Amer. J. Public Health* 89 (1999): 1151-53.

[21] Carl Seltzer, "Conflicts of Interest" and "Political Science," *J. Clin. Epidemiology* 50 (1997): 627-29.

[22] Simon Chapman, "Advocacy in Public Health: Roles and Challenges." *Int'l J. of Epidemiology* 30 (2001): 1226-32.

[23] Andrew C. Revkin, "Panel of Experts Faults Bush Plan to Study Climate." *N.Y. Times.* (Feb. 26, 2003): at A1.

[24] Linda Rosenstock & Lore Jackson Lee, "Attacks on Science: The Risks to Evidence-Based Policy." *Amer. J. Public Health* 92 (2002): 14-18. .

[25] Callahan, *op. cit.*

[26] Deborah Barnes & Lisa Bero. "Industry-funded Research and Conflict of Interest: An Analysis of Research Sponsored by the Tobacco Industry through the Center for Indoor Air Research." *J. of Health Policy, Politics & Law* 21 (1996): 515-42.

[27] Elisa Ong & Stanton Glantz. "Constructing 'Sound Science' and 'Good Epidemiology': Tobacco, Lawyers and Public Relations Firms." *Amer. J. Public Health* 91 (2001):1749-57.

[28] Deborah Barnes & Lisa Bero. "Why review articles on the health effects of passive smoking reach different conclusions." *JAMA* 279 (1998): 1566.

[29] Richard Deyo, Bruce Psaty, Gregory Simon, Edward Wagner & Gilbert Omenn, "The Messenger Under Attack – Intimidation of Researchers by Special-Interest Groups." *N. Engl. J. Med.* 336 (1997): 1176-80.

[30] "Clinton refuses to fund needle-exchange program." *AIDS Policy & Law* (May 1, 1998): at 1.

[31] Linda Rosenstock & Lore Jackson Lee, "Attacks on Science: The Risks to Evidence-Based Policy." *Amer. J. Public Health* 92 (2002):14 .

[32] *ibid.*

[33] Chris Collins & Thomas Coates, "Science and Health Policy: Can They Cohabit or Should They Divorce?" *Amer. J. Public Health* 90 (2000): 1389-90.

ABOUT THE CONTRIBUTORS

DERYCK BEYLEVELD—is Professor of Jurisprudence at the University of Sheffield and Director of the Sheffield Institute of Biotechnological Law and Ethics (SIBLE). His publications span criminology, legal and moral philosophy, contract law, product liability law, international human rights law, bioethics and biolaw, and public perception of biotechnology. They include: *A Bibliography on General Deterrence Research; The Dialectical Necessity of Morality; Law as a Moral Judgment; Mice, Morality; and Patents;* and *Human Dignity in Bioethics and Biolaw* (the latter three with Roger Brownsword—with whom he is currently writing a book on *Consent in the Law* for Prentice Hall).

MICHAEL BOYLAN—(Ph.D. University of Chicago) is Professor of Philosophy at Marymount University of Arlington, VA. He is the author of *Basic Ethics* (2000) an essay on normative and applied ethics, with Kevin E. Brown, *Genetic Engineering: Science and Ethics on the New Frontier* (2002), with James A. Donahue, *Ethics Across the Curriculum: A Practice-Based Approach* (2003) and forthcoming *A Just Society* (2004). In addition he has published over 60 articles covering topics in philosophy and in literature.

DAVID CUMMISKEY—is an Associate Professor of Philosophy at Bates College. In addition to his interest in medical ethics, he also works on Kantian ethics and consequentialism, *Kant's Consequentialism* (Oxford, 1996). Currently, he is writing a book, *Global Healthcare*, for Prentice Hall.

MICHAEL GREEN— Michael Green is Assistant Professor of Philosophy and a member of the Human Rights Program at the University of Chicago. He works on ethics, political philosophy, and early modern philosophy. His publications include "Justice and Law in Hobbes," *Oxford Studies in Early Modern Philosophy* (forthcoming), "Institutional Responsibility for Global Problems," *Philosophical Topics* (Fall, 2002), "The Idea of a Momentary Self and Hume's Theory of Personal Identity," *British Journal for the History of Philosophy* (1999), and "National Identity and Liberal Political Philosophy," *Ethics and International Affairs* 10 (1996).

D. MICAH HESTER—is Assistant Professor of Biomedical Ethics and Humanities at Mercer University School of Medicine. His scholarship has focused on the ethics of physician-patient relationships and the accounts of human nature upon which they are founded. He is author of *Community as Healing: Pragmatist Ethics in Medical Encounters* (2001) and co-author (with Rob Talisse) of *On James* (2003). His most recent journal publications can be found in the *Cambridge Quarterly of Healthcare Ethics* and the *Journal of Medicine and Philosophy.*

JACQUELYN KEGLEY— is Professor of Philosophy at CSU, Bakersfield and a recipient of the Wang Family Excellence award for Outstanding Teaching, Research, and Service. Author*, Genuine Individuals and Genuine Communities. A Roycean Public Philosophy* (Vanderbilt University Press, 1997) and of over 45 articles in American philosophy, bioethics and technology. She is Lifetime Senior Fellow and founder of an Institute of Ethics at California State University, Bakersfield and also a member of a UNESCO sponsored group that is developing Case Studies and educational materials in ethics for use internationally in medical schools.

MARY BRIODY MAHOWALD— PhD (philosophy) is Professor Emerita in the College, MacLean Center for Clinical Medical Ethics, and Department of Obstetrics and Gynecology at the University of Chicago. In addition to over 100 articles in philosophy and clinical journals, her recent books include *Women and Children in Health Care: An Unequal Majority* (Oxford, 1993), *Disability, Difference, Discrimination: Perspectives on Justice in Bioethics and Public Policy* (co-authored with Anita Silvers and David Wasserman; Rowman & Littlefield, 1998), and *Genes, Women Equality* (Oxford, 2000). Currently, she is Visiting Professor Emerita at Stanford University, where she is completing a book about ethical issues in women's health care across their lifespan.

SHAUN PATTINSON— is a Lecturer in the Faculty of Law at the University of Sheffield. His research interests are primarily in the area of medical law and ethics, especially the genetic and reproductive techniques on which he has recently published a book entitled *Influencing Traits Before Birth.*

EDMUND D. PELLEGRINO, M.D., M.A.C.P.—is the Professor Emeritus of Medicine and Medical Ethics at the Center for Clinical Bioethics at Georgetown University. He is a Master of the American College of Physicians and was the John Carroll Professor of Medicine and Medical Ethics and the former director of the Kennedy Institute of

Ethics at Georgetown University. During Dr. Pellegrino's career in medicine and university administration, he has been a departmental chairman, dean, vice chancellor, and president and has published over five hundred items in medical science, philosophy, and ethics. His interests include the history and philosophy of medicine, professional ethics, and the physician-patient relationship. Professor Pellegrino has been a leading developer of the academic study of biomedical ethics. He is co-author (along with David C. Thomasma) of *A Philosophical Basis of Medical Practice, Toward a Philosophy and Ethic of the Healing Professions* (Oxford: Oxford Univ. Press, 1981) and *For the Patient's Good: The Restoration of Beneficence in Health Care* (Oxford: Oxford University Press, 1987).

LAURA PURDY—teaches in the Public Affairs, Ethics, Law, and Social Policy major, and in Women's Studies at Wells College. She has also taught at Cornell University, Hamilton College, and served as a bioethcist at a major teaching hospital affiliated with the University of Toronto. She is the author (or co-author) of six books and numerous articles in bioethics, reproduction, feminism, education, and family issues. She hopes to leave the world a better place than she found it, and would like to think that her writing and teaching has in some way contributed to that goal.

ROSEMARY QUIGLEY—is Assistant Professor of Medical Ethics and health policy at Baylor College of Medicine in Houston, Texas. She is a Greenwall Faculty Scholar in Bioethics and Health Policy and her research concentrates on research ethics, genetics, and disability. Quigley served on the NIH Director's Council of Public Representatives from 1999-2003. She graduated from Harvard College, where she studied English literature, and received her J.D. and M.P.H. (health management and policy) from the University of Michigan.

ROSAMOND RHODES, PH.D.—is Professor of Medical Education and Director of Bioethics Education at Mount Sinai School of Medicine and a member of the doctoral faculty in philosophy at The Graduate School, CUNY. She is co-editor of the American Philosophical Association Newsletter on *Philosophy and Medicine* and co-editor of two collections of essays: *Physician Assisted Suicide: Expanding the Debate* (Routledge, 1998), and *Medicine and Social Justice: Essays on the Distribution of Health Care* (Oxford, 2002). Her most recent work has been focused on the application of constructivist ethics and political philosophy to create an understanding of professional responsibility and other contentious issues in bioethics.

WANDA TEAYS— is Professor of Philosophy and department chair at Mount St. Mary's College in Los Angeles. She is the author of *Second Thoughts: Critical Thinking for a Diverse Society.* 2nd ed. (McGraw Hill, 2003), co-editor (with Laura Purdy) of *Bioethics, Justice, and Health Care* (Wadsworth, 2001), co-editor (with Stanley G. French and Laura M. Purdy) of *Violence Against Women: Philosophical Perspectives* (Cornell University Press, 1998), and author of articles in bioethics, social justice, and the law.

DAVID C. THOMASMA—Before his death, Dr. Thomasma was professor and Fr. English Chair in Bioethics at the Neiswanger Institute for Bioethics and Health Policy at the Loyola University of Chicago. He has been a pioneer in the academic study of bioethics. He was also co-author (with Edmund D. Pellegrino) of *A Philosophical Basis of Medical Practice, Toward a Philosophy and Ethic of the Healing Professions* (Oxford: Oxford Univ. Press, 1981) and *For the Patient's Good: The Restoration of Beneficence in Health Care* (Oxford: Oxford University Press, 1987). It was David Thomasma who accepted this book into his series with Kluwer and it is to him that this book is dedicated.

ROSEMARIE TONG— is Distinguished Professor of Health Care Ethics in the Department of Philosophy and Director of the Center for Professional and Applied Ethics at the University of North Carolina at Charlotte. Dr. Tong is the author of several publications and she has given addresses on a variety of topics, including bioethics, aging and long term health care issues, women's health issues, death and dying, and genetic engineering.

She is a member of several health care ethics committees, institutional review boards, and editorial boards. She is a Board Member of the Bioethics Resource Group, a grassroots citizens group in North Carolina dedicated to increasing the public's knowledge about health care issues, policies, and problems. Her interests are genetic and reproductive technology, eliminating national and global healthcare disparities, complementary and alternative medicine, and issues related to aging well. Currently, she is writing a book on Health Care Ethics for Allied Health Professionals.

INDEX

International Library of Ethics, Law, and the New Medicine

1. L. Nordenfelt: *Action, Ability and Health*. Essays in the Philosophy of Action and Welfare. 2000   ISBN 0-7923-6206-3
2. J. Bergsma and D.C. Thomasma: *Autonomy and Clinical Medicine*. Renewing the Health Professional Relation with the Patient. 2000   ISBN 0-7923-6207-1
3. S. Rinken: *The AIDS Crisis and the Modern Self*. Biographical Self-Construction in the Awareness of Finitude. 2000   ISBN 0-7923-6371-X
4. M. Verweij: *Preventive Medicine Between Obligation and Aspiration*. 2000
                                                                        ISBN 0-7923-6691-3
5. F. Svenaeus: *The Hermeneutics of Medicine and the Phenomenology of Health*. Steps Towards a Philosophy of Medical Practice. 2001   ISBN 0-7923-6757-X
6. D.M. Vukadinovich and S.L. Krinsky: *Ethics and Law in Modern Medicine*. Hypothetical Case Studies. 2001   ISBN 1-4020-0088-X
7. D.C. Thomasma, D.N. Weisstub and C. Hervé (eds.): *Personhood and Health Care*. 2001
                                                                        ISBN 1-4020-0098-7
8. H. ten Have and B. Gordijn (eds.): *Bioethics in a European Perspective*. 2001
                                                                        ISBN 1-4020-0126-6
9. P.-A. Tengland: *Mental Health*. A Philosophical Analysis. 2001   ISBN 1-4020-0179-7
10. D.N. Weisstub, D.C. Thomasma, S. Gauthier and G.F. Tomossy (eds.) : *Aging: Culture, Health, and Social Change*. 2001   ISBN 1-4020-0180-0
11. D.N. Weisstub, D.C. Thomasma, S. Gauthier and G.F. Tomossy (eds.) : *Aging: Caring for our Elders*. 2001   ISBN 1-4020-0181-9
12. D.N. Weisstub, D.C. Thomasma, S. Gauthier and G.F. Tomossy (eds.) : *Aging: Decisions at the End of Life*. 2001   ISBN 1-4020-0182-7
                                          (Set ISBN for Vols. 10-12: 1-4020-0183-5)
13. M.J. Commers: *Determinants of Health: Theory, Understanding, Portrayal, Policy*. 2002
                                                                        ISBN 1-4020-0809-0
14. I.N. Olver: *Is Death Ever Preferable to Life?* 2002   ISBN 1-4020-1029-X
15. C. Kopp: *The New Era of AIDS*. HIV and Medicine in Times of Transition. 2003
                                                                        ISBN 1-4020-1048-6
16. R.L. Sturman: *Six Lives in Jerusalem*. End-of-Life Decisions in Jerusalem - Cultural, Medical, Ethical and Legal Considerations. 2003   ISBN 1-4020-1725-1
17. D.C. Wertz and J.C. Fletcher: *Genetics and Ethics in Global Perspective*. 2004
                                                                        ISBN 1-4020-1768-5
18. J.B.R. Gaie: *The Ethics of Medical Involvement in Capital Punishment*. A Philosophical Discussion. 2004   ISBN 1-4020-1764-2
19. M. Boylan (ed.): *Public Health Policy and Ethics*. 2004
                                          ISBN 1-4020-1762-6; Pb 1-4020-1763-4
20. R. Cohen-Almagor: *Euthanasia in the Netherlands*. The Policy and Practice of Mercy Killing. 2004   ISBN 1-4020-2250-6

KLUWER ACADEMIC PUBLISHERS – DORDRECHT / BOSTON / LONDON